AF379234

WHAT WERE THEY THINKING?

IS POPULATION ECOLOGY
A SCIENCE?

Papers, Critiques, Rebuttals,
and Philosophy

by

Bertram G. Murray, Jr.

Copyright © 2011 by Bertram G. Murray, Jr.
Library of Congress Control Number: 2010942849

ISBN 978-0-7414-6393-7 Paperback
ISBN 978-0-7414-6394-4 Hardcover
ISBN 978-0-7414-9325-5 eBook

Printed in the United States of America

Published April 2014

INFINITY PUBLISHING

Toll-free (877) BUY BOOK
Local Phone (610) 941-9999
Fax (610) 941-9959
Info@buybooksontheweb.com
www.buybooksontheweb.com

Table of Contents

Editors' Preface

Do not go gentle into that good night— Dylan Thomas

Bert Murray was working on two books when he died on 8 August 2010. This one, which he asked us to see to completion, deals with the philosophy of science, his unceasing search for patterns and laws in biology, and critiques of the state of biology, as revealed in his unpublished manuscripts and expansive letters to editors.

In reading the manuscript we could not help being impressed, as we have been for many decades, with his original and unique approach to scientific problems. Name another ornithologist who would even dream of writing about the population dynamics of the Ivory-billed Woodpecker--a species for which there is no information. Excited about its "rediscovery" and possibility of seeing one, Bert used his knowledge about the dynamics of small populations to consider the odds. Following sound scientific procedures, he looked at the specifics–i.e., similar species for which information does exist–then extrapolated to reach a conclusion. That it did not support the idea of a remnant population (or even an individual) did not endear him to those entrusted with developing "recovery plans" for something that might not exist, and several editors turned the paper down. It would have been among the interesting and relevant papers to appear in any recent ornithological journal.

Bert worked under the handicap of being an original thinker and, therefore, an iconoclast in the best sense of that word. In this book you will learn that the most cited paper in ornithology since 1975 (and maybe ever) is mathematically incorrect. Measuring clutch success, an essential parameter for understanding bird populations, requires data on the persistence of clutches from start to finish, which is very hard to obtain. Harold Mayfield devised a widely-adopted solution. But when Bert decided to test it by using an imaginary population for which all relevant data were *known,* he got different answers. No matter, his critique he was rejected by a half-dozen journals on the advice of perhaps 12 reviewers who (as experts chosen by the various editors) had almost certainly adopted the Mayfield Method and published accordingly. The reasons for rejecting Bert's critique were many, running the gamut from "this can't be true," to "everyone knows Mayfield has flaws and we have corrected them" to "who cares." Yet, despite this apparent knowledge of flaws (including a symposium on the subject), the rate at which Mayfield's paper is cited has barely abated in three decades. Clearly, it is hard to convince a reviewer of something when he has already published the opposite in an a refereed journal.

Over his 50 year career, Bert was interested in many problems, but he concentrated the last half of his career on studies of population biology, which demands a knowledge of life tables." His views on why versions found (and copied) in widely used ecology texts are incorrect, and how they can be improved, are essential reading.

More than anything else, Bert was interested in ideas. If his was wrong he wanted to know why. That quest was mainly in vain. His letters to editors (only rarely answered) made no attempt to conceal his frustration with peers who, in his view, could neither read nor do basic mathematics, not to mention the adversarial nature of the peer review process– i.e., judgments were reached from on high by reviewers who did not have the courage to sign their comments, and authors were not given a chance to respond. Some of his classic responses are included.

A sad aspect of our task was to realize that his

"second" book was not fully ready for publication, and that his files contain important ideas that remain to be fleshed out, along with other papers that have (of course) been rejected. Among these is a theory on the evolution of polyandry.

Joanna Burger and Joseph R. Jehl, Jr.

1 December 2010, Somerset, NJ

Author's Preface

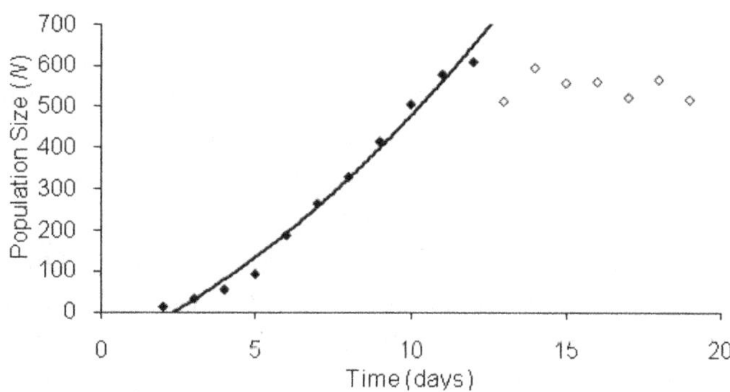

Do you think that the above graph represents an example of logistic growth? If you are an ecologist, I think you must answer "yes" because the data (from Gause 1934, *The struggle for existence*) are always presented in textbooks with a "fitted" logistic equation, showing a symmetric *S*-shaped curve with an inflection point at 0.5 K, halfway between the minimum and "mean" or "equilibrium" population size. They are shown in no other way. The follow-up question, not usually asked, much less answered, is, *why* do you believe that these data on *Paramecium aurata* represent logistic growth?

This is the subject of this book. Why do ecologists and evolutionary biologists (EEBs) believe that this graph represents logistic growth? Or, more generally, why do EEBs believe what they believe.

A great myth of ecological theory is that the logistic equation is a suitable model for describing the growth of populations. In fact, there is little empirical evidence in support of this hypothesis. I have replotted the few data (time-series counts or estimates of population size) that I have found in the literature for which logistic curves were "fitted." Plots between near minimum size and maximum size, between which logistic growth should occur, did not resemble *S*-shaped curves at all. Populations essentially grew continuously upward (*J*-shaped), as in the above graph) until the maximum size was reached, after which the populations fluctuated between an upper and lower boundary or declined. I thought this should be of interest to theoretical population ecologists, so I wrote a paper for publication. My paper began as a commentary on a paper by B.-E. Saether, S. Engen, R. Lande, C. Both, and M. E. Visser (*Oikos* 2002) on the alleged logistic growth of a Pied Flycatcher (*Ficedula hypoleuca*) population. I repeat the question: How do we know that population growth in this species is logistic? My paper was rejected in 2003. Since then, the paper has undergone revisions, expansions, and title changes while being submitted to and rejected by the editors of *Oikos* (2002), *Oecologia* (2003), *Theoretical Population Biology* (2004), *Annales Zoologici Fennici* (2004), *Austral Ecology* (2005), and *Population Ecology* (2006). I have also sent the ms. to several prominent population ecologists, none of whom even acknowledged having received it. Chapter 2 of this book is the latest manuscript.

Unfortunately, in order for my readers to understand my thinking on this, I am going to have to provide, in Chapter 1, a "crash course" regarding my thinking about the practice of science in general because my colleagues and I differ considerably in what we accept as our background knowledge. For example, I take physics to be my model of what a scientific theory should be (Murray 2001, *Biol. Rev. 76:255-289*). This is contrary to the starting point of many of my colleagues who contend that the biological world is just too complex, especially when compared with the physical world, to accommodate universal laws from which predictions may be deduced.

Physics vs. Biology

Physics is characterized by laws and predictive theory. Laws are statements that are assumed to be universally true. Predictions are statements that are logically deduced from the laws and the initial conditions of the experiment or situation. The most famous and probably the easiest to understand is Newton's Theory of Motion. Newton proposed three laws, including the First Law of Motion, which states, "Every body continues in its state of rest, or of uniform motion in a right [straight] line, unless it is compelled to change that state by forces impressed upon it". This law is an invention of Newton's mind. There is no observation in the universe that even suggests that a body in motion should remain in motion forever until acted upon by an outside force. Newton had no justification for asserting that the First Law was universally true. He stated it as an axiom—a statement that he *assumed* to be universally true— of his theory. Nevertheless, with it, he worked out the "Moon problem." Using his laws and his knowledge of the diameter and period of the Moon's orbit (the initial conditions), he accurately predicted the acceleration of a body near the Earth's surface, which could be measured. He and other physicists used his laws and other sets of initial conditions to predict or otherwise explain a wide range of physical phenomena, such as the shape of the earth (i.e. flattened at its poles), the daily cycle of two high and two low oceanic tides, the precession of the equinoxes, the orbits of comets, the position of an unknown planet, and much more.

Ecologists and evolutionary biologists (EEBs) do not have laws and predictive theory, and even argue that "laws" can play *no* role in theoretical biology (See Chapter 1). As O'Hara (2005, *Oikos 110:390-393*) snidely stated in the title of his peer-reviewed paper, "we don't need no stinkin' laws." Why should EEBs and physicists have such different approaches toward theoretical science? I think it is a matter of their training. As Dorothy Sayer's Lord Peter Wimsey put it: "But you see, I can believe a thing without understanding it. It's all a matter of training." If this is so, we may ask, How are biologists trained regarding "theory." The answer is given at the beginning of almost every biology textbook in a section often entitled, "The scientific method." I happen to have at hand an old text, the second edition of *Biological Science* Keeton (1972) starts:

Science cannot deal with anything that cannot be observed.

I wonder what Keeton meant by "observed." As far as I know, protons, electrons, strong forces, weak forces, inertia, gravitation, and magnetism cannot be directly observed. What we observe are the effects of these hypothetical particles and processes. Keeton continues:

Once a scientist has made careful observations, he must do something with his data. It is not enough simply to amass data; the data must be fitted into some sort of generalization. The observation that a particular fly has three pairs of legs is interesting, but it will remain an essentially useless isolated fact until it is incorporated into a generalization such as, "All flies have three pairs of legs," or the even broader generalization, "All insects have three pairs of legs." The step from isolated observations to generalization can be made with confidence only if enough observations have been made to give a firm basis for generalization and only if the individual observations have been reliably made.

After the good scientist has reasoned inductively from the specific to the general (i.e., from specific observations to a general statement), he must reverse his field and reason deductively, from the general to the specific. … [Having arrived at the generalization, "All insects have three pairs of legs," he] may now reason deductively that all insects have three pairs of legs, and if cockroaches, crickets, moths, and bees are insects, then they must have three pairs of legs. In other words, he uses his generalization to formulate a hypothesis about things he has not yet observed. …

The scientist's query must always be: "What is the evidence?"

This is a wonderful description of induction. It is *not*, however, the scientific method of physicists. According to Einstein,

The truly great advances in our understanding of nature originated in a way almost diametrically opposed to induction. The intuitive grasp of the essentials of a large complex of facts leads the scientist to the postulation of a hypothetical basic law, or several such laws. From these laws, he derives his conclusions, ... which can then be compared to experience. Basic laws (axioms) and conclusions together form what is called a "theory." Every expert knows that the greatest advances in natural science ... originated in this manner, and that their basis has this hypothetical character.

A good example of physicists' thinking is Isaac Newton, who began (page 14 of the 1952 reprint) his theoretical argument with his First Law. A well-trained biologist would ask Newton, "What is your evidence for the First Law?"; to which Newton would raise his eyebrows and reply, "There is no evidence. There is not any conceivable way of observing inertial motion. That is why I have asked you to *believe* that the First Law is universally true." The biologist thinks for a minute. "Why should anyone else believe it is true?" Newton replies, "Well, with my three laws and our knowledge of the period and radius of the Moon's orbit, we can predict the acceleration of a body near the Earth's surface, a prediction that is remarkably close to the observed value. And with the laws and different sets of facts (initial conditions), we can predict a whole lot of other things. You should read my book." The good biologist thinks more deeply (remember, he is living at the turn of the 17th Century). He asks, "If they cannot be observed, where did the laws come from in the first place? How did you make the connection between the Moon's motions and an apple falling from a tree?" "Ah," Newton answers, "It's all in the mind. It's conceptual. I visualize what the world looks like and how it works. The real problem is trying to translate that vision into words, equations, and testable predictions so that others can see my vision."

In contrast, Keeton's biologist concludes from his observations on insects, "All insects have three pairs of legs." His prediction that all insects that have not yet been discovered will have three pairs of legs does not seem to be of the same level of intellectual accomplishment as Newton's prediction of a falling body's acceleration on Earth from his knowledge of the motions of the Moon and his three Laws, which we may recognize as *bold guesses* (following Feynman [1965,*The character of physical law*] and Popper [1968 *The logic of scientific discovery*]). The biologists' generalization about insects is certainly not a bold guess. Indeed, it is indistinguishable from the argument, "all the swans I have seen in my life have been white; therefore, all swans are white." The latter argument has been discussed so often by philosophers of science that its form should be instantly recognized.

Einstein wrote "The grand aim of all science is to cover the greatest number of empirical facts by logical deduction from the smallest number of hypotheses or axioms." For reasons that I cannot explain, I think that I have always subscribed to that view. It seemed natural that I should be able to deduce my current observations from some sort of overriding principles. It seemed trivial to me to "deduce" that the next insect I saw would have three pairs of legs because all the insects I have seen so far have had three pairs of legs. So, I formalized this approach to explanation with the question, "What statements can I make from which I can deduce my data?" My theory of evolution, thus, starts with three laws (Murray, *Population Dynamics: Alternative Models*, 1979; *Proc. 22 Int. Ornithol. Congr.*, 1999; *Oikos*, 2000; *Biol. Rev.*, 2001). The deductions that I made from these three laws and appropriate initial conditions include (1) that the size of clutches of birds should increase with increasing latitude, (2) that clutch size should increase, not decrease, with increasing mortality of eggs and chicks in the nest (contrary to the conventional wisdom), and (3) that clutch size should become smaller as life expectancy gets longer. All of these predictions seem consistent with the empirical facts. Furthermore, from these same ideas, I predicted that the sex ratio of the polyandrous Spotted Sandpiper *Actitis macularia* should be >1, when Lew Oring, world's authority on this species, believed that it was one.

Also, using these ideas, Jehl and Murray (1986, *Current Ornithology 3:1-86*) explained the reversed sexual size dimorphism found in birds. My theory of evolution seems to cover the greatest number of empirical facts by logical deduction from the smallest number of hypotheses or axioms (only three compared with the myriad of ad hoc hypotheses currently required to explain life history traits). Yet, my colleagues have completely ignored my theory: "What were they thinking?" I wonder.

This book is about my speculations regarding what my colleagues have been thinking. Our approaches toward scientific research are strikingly different. The explanations of ecologists and evolutionary biologists are essentially ad hoc hypotheses, whereas my explanations take the form of deductive-nomological theories. I have been interested in developing a Newtonian-like universal theory, characterized by a small number of laws from which the many facts of biological diversity can be deduced. That sounds extraordinarily ambitious, but that is the goal I have set for myself.

I began in 1979 with several statements that I assumed were universally true. These were formalized into three "laws" of evolution and two "laws" of population dynamics and an equation for calculating the mean clutch size from other empirically determined demographic data (number of breeding seasons [determined from age-specific survival rates and age of first breeding] and number of broods reared per breeding season). Given different initial conditions (starting points) I have been able to predict the clutch sizes of two species of birds, Prairie Warbler *Dendroica discolor* and Florida Scrub-Jay *Aphelocoma coerulescens*. Dale Kennedy has done the same for the House Wren *Troglodytes aedon*. The predicted values were close to the empirical values. For example, the predicted clutch size for the Florida Scrub-Jay was 3.43 eggs, which is barely distinguishable from the actual clutch size, 3.33. To my knowledge, no other hypothesis predicts the clutch size of any species of bird. I guess I was naïve, but I thought that ornithologists with good quality data from long-term population studies would follow-up to see whether my equation was more generally applicable. So far, no one has. I wonder why. My theory and equations seem to represent a small number of assumptions from which a broad range of empirical facts can and have been deduced.

This book is also about the process of publication, and the high-jacking of scientific thinking by a few well-entrenched ideas and theories. Most of the papers in this book have been rejected, often by a number of different "reputable" journals. The reasons for the rejections have left me puzzled. The evidence for my being puzzled comes in the form of the peer reviews that I have received on my manuscripts, submitted throughout most of my career. Peter Medawar (*The Art of the Soluble*, 1967: 151) wrote, "What scientists *do* has never been the subject of scientific, that is, an ethological enquiry… It is no use looking to scientific 'papers,' for they not merely conceal but actively misrepresent the reasoning that goes into the work they describe…. Only unstudied evidence will do—and that means listening at a keyhole." Listening at keyholes is not an efficient method for finding out what scientists do and think. This is best revealed in their anonymous reviews of unpublished manuscripts. Believing that no one, other than the editor, will know their identity, reviewers can write the most outrageous nonsense.

Editors avoid their own accountability by not revealing the identity of their reviewers, that is, they do not have to defend their choice of reviewers. As Jim Lloyd (1985, *Florida Entomologist 68:134-139*) pointed out, "our review system can sometimes amount to nothing more than an adversarial confrontation where the defendant is presumed guilty, has no counsel or friend in court by arrangement, and cannot face his accusers, and there are no qualifications for judges."

One of my responsibilities as a scientist is to do research—not only to discover new "knowledge" but to publish it. In this book I have included manuscripts that I have written during the past 15 years or so. For those that were rejected, I have written to the editors regarding my reasons (evidence and arguments) for thinking my critics incorrect. Some of these letters are included.

It is not my intention to embarrass anyone. I am correcting what I perceive to be errors in the reviewing process, as practiced by my colleagues. Too often, reviewers do not substantiate their allegations with facts. The reviewers should be held to the same standards of evidence and argument as the authors are. The reviewers' responsibility is to find the authors' errors of fact and they should document those errors with citations.

My intention is to advance my science by having my views about ecological theory readily available for others to judge. Unfortunately, most of the reviews were written by anonymous people, which precludes the opportunity to have a scientific discussion with those colleagues who disagree with me. Perhaps, readers of this book will explain my errors.

Acknowledgments

I am grateful to the two people who had a seminal effect on my life and career, James Baird and H. B. "Bud" Tordoff, who were first role models, then mentors, and finally life-long colleagues. Without their help I might have stayed in the intelligence corps in the military. Norm Ford and Val Nolan were friends, colleagues and coauthors on many papers. Joe Jehl was a friend and colleague from high school days, through graduate school to the present. Hardly a manuscript of mine went to a journal without Joe's comments and careful editing. We were "brothers" in our love of birds and of science.

Others have helped along the way, as friends and colleagues, including Joanna Burger, Mike Gochfeld, and Charlie and Mary Leck. I have shared many local and foreign trips with them, as well as with Chris and Paula Williams, Wade and Sharon Wander, and Ann Hoffenberg.

In the end, I entrusted the completion of this book to Joanna Burger and Joe Jehl, and hope that it will find its way into the hands of other scientists who will appreciate the struggles this book represents. Science is a process, and we all build on the work of others.

Finally, I thank my wife Patti, whose love, concern and love of nature carried me though the years. The trips we took together were some of the happiest moments of my life. One of my favorites was to Snow Hill Island to see the Emperor Penguin colony, represented by one of Patti's photos on the cover of the book. She made my work, and the writing of this book possible.

Somerset, New Jersey
1 August 2010

Chapter 1

Introduction to Philosophy

The trouble with people is not that they don't know,
but that they know so much that ain't so.—Josh Billings

According to Isaac Newton, "Science consists in discovering the frame and operations of Nature, and reducing them, as far as may be, to general rules or laws—establishing these rules by observations and experiments, and thence deducing the causes and effects of things." A later physicist suggested, "The grand aim of all science is to cover the greatest number of empirical facts by logical deduction from the smallest number of hypotheses or axioms" (Einstein, quoted in Calaprice 2005: 256). Philosopher of science, Karl Popper (1979: 191), had a similar view, ". . . the aim of science [is] to find *satisfactory explanations*, of whatever strikes us as being in need of explanation. By an *explanation* (or a causal explanation) is meant a set of statements by which one describes the state of affairs to be explained (the *explicandum*) while the others, the explanatory statements, form the 'explanation' in the narrower sense of the word (the *explicans* of the *explicandum*)."

Einstein (1954: 221) further describes the "scientific method,"

The theorist's method involves his using as his foundation general postulates or "principles" from which he can deduce conclusions. His work thus falls into two parts. He must first discover his principles and then draw the conclusions which follow from them. For the second of these tasks he receives an admirable equipment at school. If, therefore, the first of his problems has already been solved for some field or for a complex of related phenomena, he is certain of success, provided his industry and intelligence are adequate. The first of these tasks, namely, that of establishing the principles which are to serve as the starting point of his deduction, is of an entirely different nature. Here there is no method capable of being learned and systematically applied so that it leads to the goal. The scientist has to worm these general principles out of nature by perceiving in comprehensive complexes of empirical facts certain general features which permit of precise formulation.

Once this formulation is successfully accomplished, inference follows on inference, often revealing unforeseen relations which extend far beyond the province of the reality from which the principles were drawn. But as long as no principles are found on which to base the deduction, the individual empirical fact is of no use to the theorist; indeed he cannot even do anything with isolated general laws abstracted from experience. He will remain helpless in the face of separate results of empirical research, until principles which he can make the basis of deductive reasoning have revealed themselves to him.

In 1964 Richard Feynman delivered the Messenger Lectures at Cornell University—*The Character of Physical Law*. In these, Feynman (1965: 156, an edited transcript) described the scientific method a bit more casually.

In general we look for a new law by the following process. First we guess it. [In the film shown on TV of this lecture, his Cornell audience laughed, at which point he interrupted himself to say something like, "No kidding. We guess it."] Then we compute the consequences of the guess to see what would be implied if this law that we guessed is right. Then we compare the result of the computation to nature, with experiment or experience, compare it directly with observation, to see if it works. If it disagrees with experiment it is wrong. In that simple statement is the key to science. It does not make any difference how beautiful your guess is. It does not make any difference how smart you are, who made the guess, or what his name is—if it disagrees with experiment it is wrong. That is all there is to it.

Northrop (1947), a philosopher, distinguished two stages in the development of a science. First is the "natural history stage of inquiry," that is, the making and collection of observations, and the second is the "stage of deductively formulated theory." This point was elaborated by Hempel (1965:177-178),

Scientific systematization is ultimately aimed at establishing explanatory and predictive order among the bewilderingly complex "data" of our experience, the phenomena that can be "directly observed" by us. It is a remarkable fact, therefore, that the greatest advances in scientific systematization have not been accomplished by means of laws referring explicitly to *observables*, i.e., to things and events which are ascertainable by direct observation, but rather by means of laws that speak of various *hypothetical*, or *theoretical*, *entities*, i.e., presumptive objects, events, and attributes which cannot be perceived or otherwise directly observed by us.

For a fuller discussion of this point, it will be helpful to refer to the familiar rough distinction between two levels of scientific systematization: the level of *empirical generalization*, and the level of *theory formation*. The early stages in the development of a scientific discipline usually belong to the former level, which is characterized by the search for laws (of universal or statistical form) which establish connections among the directly observable aspects of the subject matter under study. The more advanced stages belong to the second level, where research is aimed at comprehensive laws, in terms of hypothetical entities, which will account for the uniformities established on the first level.

We should note that Newton, Einstein, Feynman, Popper, Northrop, and Hempel are describing the aim and methods of the *physical* sciences. These do not seem to be the aim or methods of biologists. Indeed, several prominent biologists contend a priori that biology can have no laws, that is, it can have no axioms or general principles from which conclusions or predictions may be deduced (Cohen 1971, Van Valen and Pitelka 1974, Stebbins 1977, Strong 1980, Bartholomew 1982, Mayr 1982, Quinn and Dunham 1983, Roughgarden 1983, Strong 1983, Bartholomew 1986, Egler 1986, Slobodkin 1988, Lawton 1999). According to them, the biological world is just too complex, especially when compared with the physical world.

Bartholomew (1986) contended,

Not only are biological systems staggeringly complex, they have an evolutionary history shaped by chance and natural selection, which is both stochastic and blind to future consequences. The consequences of natural selection require biologists to use procedures that sometimes differ from those used by physical scientists.

Bartholomew (1986) believed also that philosophers of science have misled biologists into believing that "we should use the same procedures that have worked so well in physics and chemistry," particularly "deductive logic," and that "Natural history tells us unequivocally that we are foolish to look for general answers to specific questions about how organisms perform." From my experience these biologists represent the feelings of biologists in general with regard to the usefulness of deductive logic or the methods of physicists in solving biological problems. I disagree (see Box 1). In my view, biologists have failed to establish general principles *not* because they have emulated physicists (as explicitly suggested by Cohen [1971], Bartholomew [1986], and Egler [1986]) but because they *have not* emulated physicists.

What exactly were Newton, Einstein, Feynman, Popper, Northrop, and Hempel referring to when they suggested that scientists should be searching for theories that explained the greatest number of empirical facts by logical deduction from the smallest number of hypotheses or axioms? An explicit description has been provided by Hempel and Oppenheim (1948), which is called the **deductive-nomological** (D-N) method, sometimes known as the Popper-Hempel model. This method is virtually unknown to biologists, so it is worth detailed discussion.

Deductive-Nomological Theory

In outline, the D-N model comprises three parts, (1) Laws; (2) Initial conditions; and (3) Predictions, and it is best illustrated with a scientific example. Philosophers and biologists *usually* discuss models in the abstract, but models and methods are better understood with a concrete example. The example that I am going to use is the venerable theory of Isaac Newton.

Laws

The laws are statements that are assumed to be universally true. I say "assumed" because we can *never* know that any scientific statement is universally true. Furthermore, to paraphrase Einstein, "there is no method capable of being learned and systematically applied so that it leads to the goal" of discovering universally applicable laws. To quote Einstein again, "The scientist has to worm these general principles out of nature by perceiving in comprehensive complexes of empirical facts certain general features which permit of precise formulation." Feynman says simply that we must "guess" them.

Laws are guesses, assumptions, conjectures, hypotheses, rules, and principles. They are made up out of a scientist's imagination, and with them the scientist hopes to explain a segment of the observable world by predicting it. The difference between "laws" and "guesses" etc. is that when we label a statement a "law," we assume that our guess, conjecture, hypothesis, rule, or principle is not only universally true but *necessary* for making logically deduced predictions. The reason that we assume a law expresses a universal truth is to prevent a scientist's weaseling out of discrepancies between the predictions of theory and the empirical facts. He is not allowed to say, "My theory does not work in this situation because my law does not apply to this situation."

Sometimes this is explicit. For example, Lack (1947, 1954, 1966) proposed that the mean clutch size of birds was limited by the average amount of food available for producing eggs by the laying females: the "Egg Production Hypothesis" (EPH). The idea is that birds at higher latitudes during the breeding season have longer days in which to gather food and, thus, could lay larger clutches than tropical species. This hypothesis has been challenged frequently as a general explanation for the evolution of clutch size (e.g., Skutch 1949, Cody 1966, Murray 1979). One of the challengers was the "Egg Viability Hypothesis," which was proposed explicitly to explain clutch-size variations in waterfowl (Arnold et al. 1987). Furthermore, according to them, the EPH could not explain the variations in waterfowl. Nevertheless, Ankney, Afton, and Alisauskas (1991) rejected the egg viability hypothesis because "We are unaware of data from waterfowl that are inconsistent with the 'Egg Production Hypothesis.' Thus, we urge critics of the EPH to attempt to refute it in the 'the old fashioned way.' That is, obtain data from waterfowl; according to the AOU checklist, coots do not qualify!" The implication is that data obtained from a study of the American Coot (*Fulica americana*) cannot be used to test a hypothesis designed to explain clutch-size variations in other birds. Explicitly, the EPH is not a universal explanation.

When we label a statement as a guess, conjecture, hypothesis, rule, or principle, however, we do not imply that it is universally true. Such statements may be useful generalizations, but they are not testable, more specifically, not falsifiable. Bergmann's Rule, for example, is not refuted simply because we have found a species that contradicts it. Nevertheless, it is useful for us to know it.

Newton's First Law of Motion is, "a body at rest will remain at rest and a body in motion will remain in motion at a constant speed in a straight line indefinitely, unless acted on by an outside force." There is no way that this statement could be inferred from direct observation of the motions of bodies in the universe, if only because no bodies in the universe are moving in straight lines. It had to be guessed. Furthermore, it could not be tested directly. Einstein and Infeld (1938: 8) pointed out, "We have seen that this law of inertia cannot be derived directly from experiment, but only by speculative thinking consistent with observation. The idealized experiment can never be actually performed, although it leads to a profound understanding of real experiments." (For further discussion of the meaning of "law," see below.)

Initial conditions

Initial conditions are usually statements that are known to be true. Many of these are the facts of nature, such as the radius and period of the orbit of the Moon. Many others may be the conditions that we set up in a laboratory experiment. If such-and-such is true in nature, or if such-and-such are the conditions of our experimental set-up, what then should we expect to see if the laws that we assumed to be true are true?

Predictions

Predictions are statements that are logically deduced from the application of the laws to the situation described by the initial conditions. A scientific theory is tested by comparing its predictions with the empirical facts.

The D-N model may best be illustrated with the example of Newton's "Moon problem." Newton very clearly described the train of thought from his laws to his predictions, and the story has been told many times (e.g., Rogers 1960, Cohen 1980), albeit in somewhat different ways. The first thing that Newton had to do was to establish a connection between Galileo's terrestrial mechanics (the motion of cannonballs) and Kepler's celestial mechanics (the motion of planets). The story goes that Newton was thinking about apples. Suppose the force that attracted apples toward the earth, the gravitational force, extended out to the moon. What would happen? The moon should fall to the earth, but it does not. Thus, Newton reasoned, some countervailing force must be acting on the moon, tending to move the moon away from the earth. The moon's orbit, perhaps, was a consequence of these two forces.

Newton was also aware that if he threw an apple, the apple would travel farther if he applied a greater force to it. He thought that if he threw an apple with sufficient force, it would circle the earth just as planets do. Thus, the moon does not fall to the earth, as a dropped apple does, because another force, called the centrifugal force is counteracting the gravitational force. Furthermore, if the gravitational force were not acting on the apple, the thrown apple would move in a straight line *away* from the earth. It seemed possible, then, that the orbit of the moon was a consequence of two interacting forces, the gravitational force drawing the moon toward the earth and a centrifugal force taking the moon away from the earth. Great idea. Popper would no doubt call it a bold conjecture. The problem, of course, was to prove it.

Newton imagined his three laws of motion and his two propositions regarding the gravitational force between two bodies—that it was directly proportional to the product of their masses and indirectly proportional to the square of the distance between them. This is the first step, for which there are no rules that guarantee success. A scientist can only imagine in his brain the rules that run the world and try them out.

Newton conjectured that the magnitude of the gravitational force that is pulling the apples and the moon toward the earth varies as the inverse of the square of their distance from the earth. The apple is one earth radius from the earth's center, and the moon is 60 earth radii from the earth's center. Thus, the gravitational force on an apple should be 60^2 times greater than that on the moon. Because the distance fallen is proportional to acceleration and acceleration is proportional to force, the distance fallen by an apple in one second should be 3,600 times greater than the distance fallen by the moon in one second. The distance fallen by an apple in one second near the earth's surface, then, should be about 16.1 feet.

The simple experiment, then, is to go to the orchard and measure the distance an apple falls in the first second of free-fall. But, alas (for the story), he did not. Newton used the experimental data of Huygens on the swing of pendulums. The confirmation of such a prediction gave physicists confidence in Newton's Laws. Furthermore, as Einstein said, "inference follows on inference, often revealing unforeseen relations which extend far beyond the province of the reality from which the principles were drawn." With his laws applied to different sets of initial conditions, Newton predicted or otherwise explained Kepler's three Laws of Planetary Motion, the relative masses of the planets, moon, and sun, the bulge of the earth at the equator, the precession of the equinoxes, the moon's motions, and the earth's tides (Rogers 1960). Newton had covered a great number of empirical facts by logical deduction from a small number of hypotheses or axioms.

Newton's theory was enormously successful. Nevertheless, no matter how successful it was, no matter how many times its predictions were confirmed by observation and experiment, Newton's Laws always remained hypothetical. All theories starting with universal laws are always hypothetical because we can never do all the imaginable experiments to test them. Furthermore, the predictions of one set of laws can be made by another set of laws. Newton's theory was finally replaced by Einstein's theory of general relativity two centuries later, not because it was "simpler" but because it covered a greater number of empirical facts. Einstein's theory explained all that Newton's did and more.

Einstein (quoted in Calaprice 2005:225) observed,

The truly great advances in our understanding of nature originated in a way almost diametrically opposed to induction. The intuitive grasp of the essentials of a large complex of facts leads the scientist to the postulation of a hypothetical basic law, or several such laws. From these laws, he derives his conclusions, ... which can then be compared to experience. Basic laws (axioms) and conclusions together form what is called a "theory." Every expert knows that the greatest advances in natural science ... originated in this manner, and that their basis has this hypothetical character.

It is doubtful that biologists know this because they do not do this—that is, they do not think deductively from laws. They do not think deductively from laws because they are taught that their subject is too complicated for universal laws and deductive theories (see below).

Mathematics

Mathematicians think in much the same way as do physicists. This has been nicely explained by Livio (2002: 188, his italics),

Euclid attempted to build an entire theory of geometry on a well-defined logical base. Accordingly, he started with only five postulates or *axioms*, assumed to hold true, and sought to prove all the other propositions on the basis of those postulates by logical deductions. Axioms are like the rules of the game, the "truth" of which is not to be disputed. If you want to change the axioms you would be playing a different game. For instance, the first axiom states, "Between any two points a straight line may be drawn." Euclidean geometry describes propositions that are deduced to be true if this and the other axioms hold.

Note that the axioms of mathematics are *assumed* to hold true. The fifth axiom of Euclidean geometry—one version of which is, "Given a line and a point not on the line, it is possible to draw exactly one line parallel to the line through that point" (Livio 2002: 188)—is true enough for Euclidean geometry. Later mathematicians questioned this assumption. By replacing the fifth axiom with, "Through a point not on a line, there is either more than one line parallel to the given line or none," mathematicians were able to develop several non-Euclidean geometries.

Ad hoc hypotheses

If the truth be told, today's ecologists and evolutionary biologists are thinking neither deductively nor inductively because they are not looking for universal statements, whether they are the intuitive guesses from which facts may be deduced or generalizations determined from accumulated observations.

Physics and biology present a striking contrast when their characteristics are compared.

Physics	Biology
Unifiers	Diversifiers
Falsificationists	Verificationists
Universal laws	Ad hoc hypotheses
Analytical	Descriptive

No doubt, many biologists would balk at being identified as verificationists because they believe they are Popperian falsificationists. They believe that Popper's instruction was for scientists to propose "testable" hypotheses, and this, of course, he did. Could he have suggested otherwise? Unfortunately, biologists have not read Popper. What Popper (1979: 193, his italics) wrote about testable hypotheses was,

Only if we require that explanations shall make use of universal statements or laws of nature (supplemented by initial conditions) can we make progress towards realizing the idea of independent, or non-*ad hoc*, explanations. For, universal laws of nature may be statements with a rich content, so that *they may be independently tested everywhere*, and at all times. Thus, if they are used as explanations, they *may* not be *ad hoc* because they *may* allow us to interpret the *explicandum* as an instance of a reproducible effect. All this is only true, however, if we confine ourselves to universal laws which are testable, that is to say, falsifiable.

This seems rather explicit. The alternative to ad hoc hypotheses is universal laws because only universal laws are falsifiable. I know of no one, other than myself, who has proposed deductive-nomological theories with explicit universal laws, initial conditions, and predictions. If any other theoretical biologist is a Popperian falsificationist, I should like to know who he or she is. I do not say this to draw attention to myself as being superior. I point this out to put what I have published and what I write in this book in a philosophical context that seems foreign to my colleagues. "Theoretical" biologists, as well as descriptive biologists, seem to be in a different philosophical paradigm from me.

First, we scientists, as everyone else, are confronted with a chaotic picture of nature. Our first job is to discover patterns and generalizations within this welter of data. For example, we have an abundance of data on the clutch and litter sizes of organisms. Eventually, someone recognizes patterns—for example, that the clutch sizes of birds seem to increase with increasing latitude. This is the first stage of theoretical research. As Einstein pointed out, "The scientist has to worm these general principles out of nature by perceiving in comprehensive complexes of empirical facts certain general features which permit of precise formulation." Once we have the generalizations, we may test them rigorously by deducing their consequences. A rigorous test requires that the generalization be universally applicable. Unfortunately, there are no rules regarding the discovery of scientific "truth," other than a commitment to do so. We must choose whether we want to be unifiers or diversifiers in interpreting the biological facts. The evidence indicates that we biologists have been trained (perhaps unwittingly) to be diversifiers.

The laws or principles are inspired guesses, whose "truth" is established by rigorous testing of the predictions that follow from them. Biologists are not today searching for universal laws or trying to develop the logic of deductive-nomological theories because they are not taught that doing so is important. They enjoy working in the field or lab describing what they find (as I do when I have the opportunity). What they find is grist for the theoretician's mill. Therefore, what they find must be described accurately because theory cannot be any better than the empirical data it is supposed to explain. Sometimes, however, theory tells us that the data are being collected incorrectly (Chapter 3). Unfortunately, it is difficult to change the ways of the empirical biologist—"This is the way we have been collecting data for decades—this is the way we will continue collecting data." I am hopeful that the biologists' peculiar attitude about the relationship between data and theory, so well described by Fagerström (1987), will change.

I think changing the philosophy of ecologists is going to be challenging. Compare the following with what I have written so far:

(1) "... as a matter of historical record—it is standard scientific practice to reject ecological data that are in conflict with established theories; ... many contemporary ecological theories are in fact retained—and rightly so—although they are demonstrably wrong; and ... it is by no means obvious that the proximate goal of theorizing should be to find the truth, hence an important role of mathematics is to provide productive lies" (Fagerström, 1987: p. 258);

(2) "... refutations of [a] theory do not necessarily disprove or invalidate it" (Fahrig, 1988: p. 129);

(3) "Once the theory has been demonstrated to hold for at least some real ecological systems, it can be considered to be proven" (Fahrig, 1988: p. 131);

(4) "A scientific discipline like biology ... needs historical explanations which do not systematically follow the methods of investigation of Newtonian physics. Contrary to Murray's [1992a] assertions [about Newton's rules of reasoning], his recommendations may serve more to channel, rather than to improve our understanding of the evolution of living things" (Quenette and Gerard, 1993: p. 363);

(5) "I suggest that the practice of ecology be defined as the interpretation of these patterns and processes by using any approach of human endeavour that generates inspiration among practitioners for further interpretation of pattern and process in nature" (Aarssen, 1997: p. 178);

(6) "... we should stop worrying about whether or not ecology is a purely empirical science and just accept that, in practice, it is not" (Aarssen, 1999: p. 375); and

(7) "Like many other authors ..., I find the narrow emphasis on hypothesis testing counterproductive" (Turchin, 1999: p. 155).

Box 1. Natural history and deductive logic

I wrote a letter to the editor of *BioScience* in response to a paper by Bartholomew (1986) about the relationship between natural history and deductive logic (Murray 1986). It so well explains my thinking that I reproduce it here:

Bartholomew's excellent article (*BioScience* 36: 324-329) clearly shows the important role that natural history can serve in contemporary biology, which emphasizes the experimental approach to understanding suborganismal levels of biological organization. Biology is likely to suffer if biologists study organs, cells, and molecules independently of the natural history of the organisms that served as donors of those organs, cells, and molecules. Alas, however, Bartholomew suggests that "Natural history tells us unequivocally that we are foolish to look for general answers to specific questions about how organisms perform" because natural selection has produced a variety of solutions (i.e., adaptations) to the same problem facing organisms living in nature (e.g., temperature stress). He points out that there are no "single answers" to the specific questions, "how do bony fish swim? How do frogs avoid drying out?" He is quite correct and would be on solid ground if he had not contrasted the biological sciences with the physical sciences and suggested that philosophers of science have misled biologists into believing that "we should use the same procedures that have worked so well in physics and chemistry," particularly "deductive logic."

I have no doubt that Bartholomew represents the views of the great majority of organismal, ecological, and evolutionary biologists, and I have no doubt that this view is in error. It results from a misunderstanding of the nature of physical theory. Inasmuch as the natural sciences have not produced a theory comparable to Newton's, let us limit discussion to that level of scientific achievement. What kind of "specific" answers does Newton's "general" theory provide? Among other things, it predicts that the shape of a moving object's orbit should be elliptical. Is this a "specific" answer? In a way, yes. The orbits are not square, or octagonal, or any other shape. And, in a way, no. There is extraordinary diversity in the "shapes" of ellipses, from the nearly circular orbit of the planet Earth to the rather elongated orbits of comets. Just looking at the movements of planets and comets, one would be hard pressed to assume that the orbits of both (or of either!) would be elliptical. It is important to understand that Newton's theory does not predict the *specific* features of elliptical orbits, e.g., the distance between foci. Also, although Newton's theory "explains" the diurnal rhythm of the tides, it does not predict the timing of the tides at any place. (Newton's theory does make specific predictions [e.g., the distance an apple will fall in one second near the earth's surface]. The point is that not all predictions are as specific.) How specific, then, must the predictions of general theory in natural history be?

I suspect that organismal, ecological, and evolutionary biologists are recording the biological equivalents of the distance between foci and the timing of the tides, and I think they expect all the predictions of their general theories to be more specific than physicists expect (or, at least, Newton expected) of theirs, and, failing to understand that, they erroneously conclude that because general theories cannot make (or have not made) the very specific predictions they want them to make, biological diversity is not susceptible to generalization. For example, is there no conceivable general theory that might establish certain energetic principles from which hibernation, estivation, daily torpor, and temporal heterothermy might be deduced? I think so, even though I have no idea of what it might be. As I have stated earlier (*BioScience* 25: 149), "The development of further generalizations that bring unity to the diversity of our biological experience is beyond possibility for those who take it as axiomatic that biological diversity is not susceptible to generalization." Those persons who eschew deductive logic and the search for generalization in biology may well be overlooking a means of discovering interesting insights into biological processes and may be impeding the maturing of biology as a science.

Terminology

Although Popper (1959) and Humpty-Dumpty (in Carroll 1960) suggest that we should not argue about the meanings of words, it is necessary for terms to be defined. For example, if I decide to define "up" as toward the center of the earth and "down" as away from the earth, then objects will fall up instead of down. If we were to develop a theory of gravitation, it would be identical to Newton's, except that gravitation would cause bodies near the earth's surface to fall up. The meanings of up and down would have no affect on the theory. So I define the words below as I will use them. You may define them differently, but you cannot define them differently to show that I am being inconsistent or otherwise wrong in terms of your alternative definitions.

Simplicity

There is a strong belief among biologists that physics is simple compared with biology (Stebbins 1977, Bartholomew 1982, Quinn and Dunham 1983, Bartholomew 1986, Slobodkin 1988, Mayr 1996, Begon 1998). That is a common justification of the biologists' failure to develop genuinely predictive theory. My first response to such claims is that I would bet that the writer or speaker had not become a biologist because physics was intellectually unchallenging.

The fact is that physics is not simple. Physics is *simplified.* According to Newton (quoted in Cohen 1980:265),

… the orbit of any one planet depends on the combined motion of all the planets, not to mention the action of all these on each other. But to consider simultaneously all these causes of motion and to define these motions by exact laws allowing of convenient calculation exceeds, unless I am mistaken, the force of the entire human intellect.

Newton was right. The "three-body" gravitational problem that he posed has not yet been solved. In order to deal with a complex world, physicists simplify it. Newton's physical models had one planet orbiting around one sun—a two body system, despite the fact that he knew (or theorized) that "the orbit of any one planet depends on the combined motion of all the planets, not to mention the action of all these on each other."

A later physicist wrote (Einstein, quoted in Calaprice 2005:222),

In regard to his subject matter … the physicist has to limit himself very severely: he must content himself with describing the most simple events that can be brought within the domain of our experience; all events of a more complex order are beyond the power of the human intellect to reconstruct with the subtle accuracy and logical perfection the theoretical physicist demands.

Another physicist describes this complexity (Feynman 1995:23),

The things with which we concern ourselves in science appear in myriad forms, and with a multitude of attributes. For example, if we stand on the shore and look at the sea, we see the water, the waves breaking, the foam, the sloshing motion of the water, the sound, the air, the winds and the clouds, the sun and the blue sky, and light; there is sand and there are rocks of various hardness and permanence, color and texture. There are animals and seaweed, hunger and disease, and the observer on the beach; there may be even happiness and thought. Any other spot in nature has a similar variety of things and influences. It is always as complicated as that, no matter where it is.

How could we possibly know, just be looking at the "myriad of forms" that surround us, that every substance is composed of eight little particles of energy? Furthermore, Feynman (1985:8) tells us,

I must clarify something: When I say that all the phenomena of the physical world can be explained by this [QED] theory, we don't really know that. Most phenomena we are familiar with involve such *tremendous* numbers of electrons that it's hard for our poor minds to follow that complexity. In such situations, we can use the theory to figure roughly what ought to happen and this *is* what happens, roughly, in those circumstances. But if we arrange in the laboratory an experiment involving just a few electrons in simple circumstances, then we can measure it very accurately, too. Whenever we do such experiments, the theory of quantum electrodynamics works very well.

On what grounds do biologists claim that the biological world is too complex for rigorous analysis? Biologists are unaware of the facts of the physical world and unaware of physical theory. How can they possibly understand the relationship between the physical world and physical theory? Biologists claim that it is hard for their poor minds to follow biological complexity. Unfortunately, as committed logical positivists, biologists refuse to simplify the systems they are studying in order to understand them better. For example, if biologists were assigned the task of finding a relationship between mass and acceleration, they would go to the top of the Leaning Tower in Pisa, drop objects of different mass, and time the fall. They would drop cannonballs, tennis balls, and ping pong balls and find that the more massive balls fall faster. What they would not have found is Galileo's free-fall equation, which indicates that acceleration is independent of mass.

The physical world must be simplified in order for it to be studied. Francis Crick (1988) describes his early experiences as a graduate student attempting to determine the structure of complex proteins with X-ray diffraction techniques. He was at the Cavendish Laboratory at Cambridge, England, working under Max Perutz. The Cavendish was headed by Sir Lawrence Bragg, winner of the Nobel Prize for his formulation of Bragg's Law for X-ray diffraction. Of this experience, Crick (1988:47, original italics) recalls,

I made some progress with the main problem but eventually became stuck. Meanwhile Bragg had independently thought about it. Whereas I had gotten bogged down, he made rapid progress. He boldly assumed that one could approximate the shape by an ellipsoid—a particularly simple type of distorted sphere. Then he looked at what little was known of the crystals of hemoglobin of other species of animal, on the assumption that all types of hemoglobin molecules were likely to have about the same shape. Moreover, he was not disturbed if the data did not *exactly* fit his model, since it was unlikely that the molecule was *exactly* an ellipsoid. In other words he made bold, simplifying assumptions; looked at as wide a range of data as possible; and was critical but not pernickety, as I had been, about the fit between his model and experimental facts. He arrived at a shape that we now know is not a bad approximation to the molecule's real shape, and he and Perutz published a paper on it. The result was not of first-class importance, if only because the method was indirect and needed confirmation by more direct methods, but it was a revelation to me as to how to do scientific research and, more important, how *not* to do it.

This is not the way for biologists. Biologists value complexity rather than simplicity. As Dyson (1988:44-45) pointed out,

Unifiers are people whose driving passion is to find general principles which will explain everything. They are happy if they can leave the universe looking a little simpler than they found it. Diversifiers are people whose passion is to explore details. ... They are happy if they leave the universe a little more complicated than they found it. ... Biology is the natural domain of diversifiers as physics is the domain of unifiers.

Why should physicists be unifiers and biologists be diversifiers? The answer is not simply that physicists love simplicity and biologists love complexity. The difference results from a deep philosophical gulf between the methods of physicists and biologists. The logical test of physical ideas is falsification, whereas that of biological ideas is verification.

Parsimony

Related to simplicity is parsimony, or, as it is sometimes referred to, Occam's (Ockham's) razor. Parsimony refers to the simplicity of the logic rather than to simplifying the observable world. This is what Einstein meant in his statement that the aim of science is to deduce the observable facts from the "smallest number of hypotheses or axioms." Biologists often think that parsimony refers to the simplest explanation, but it does not. Newtonian theory is not "simple." Einstein's theories are not "simple." These theories are characterized by having few assumptions accepted as axioms. For example, Newton was able to predict or otherwise explain much of the known physical world in the 17th Century with three relatively simple statements—his three laws of motion (his later recognized law of gravitation was a combination of two "propositions" in the *Principia*).

Tautology

The term tautology is used often by biologists, but *always* incorrectly. A tautology is a sentence that is true by virtue of its logical form (Corcoran 1995). Thus, "All A is A" is a tautology. We can insert any word for A, and the sentence will be true. In the sentence, "All cats are cats," we do not have to know the meaning of "cats" in order to know that the sentence is true. This form of a tautology is so simple to understand that I suspect that biologists do not make this mistake very often.

Another form of tautology is the sentence, "X is either *P* or not-*P*." For example, I could say, "All swans are white or not white." My sentence is true but not very informative. You do not have to know what "swan" means to know that the sentence is true. Agatha Christie (1929) used this form when her tyro detective, Tommy, was asked, "Where is the pearl?" He replied, "The pearl is in the house, or it is not in the house." Agatha Christie knew what she was doing. She was having Tommy appear as if he were on top of the case of the missing pearl, even though he had no idea of where the pearl was.

Without realizing it, biologists make this kind of error *all* of the time. Biologists, however, never explicitly write their tautologies in this simple form, and, thus, they are unaware of them. Biological "tautologies" are much more subtle, unless you, like me, are looking for them. For example, "Why are clutch sizes smaller in the tropics than at higher latitudes?" David Lack (1947, 1948) proposed that tropical parents were unable to obtain as much food during the shorter tropical day than parents at higher latitudes could in their longer day during the breeding season. Thus, they could not feed as large a brood, and, therefore, they laid smaller clutches. Alexander Skutch (1949) replied that this was not the explanation for the small clutch sizes in the tropics. Skutch proposed that in the tropics the heavy predation on nest contents (eggs and nestlings) resulted in selection for smaller clutches. For him, the clutch size reflects the limits of food availability at high latitudes and the limits imposed by predation in the tropics. We biologists are left with a hypothesis about clutch size, "Clutches are the size they are because of (P) food limitation at high latitudes and (not-P) high predation at low latitudes." We have a problem in biology, however, because the causes of biological diversity are sufficiently complex that this statement may be true. The cause of clutch size in the tropics may in fact be different from its cause at higher latitudes. Nevertheless, Lack's general hypothesis about the effects of day length becomes an ad hoc hypothesis, applicable to clutches at higher latitudes. Skutch's hypothesis is an ad hoc hypothesis applicable to clutches at lower latitudes. We are left with two (now ad-hoc) explanations for the clutch size of birds on earth. Those of us who seek generalization might at least wonder whether there is not some general explanation, not yet thought of, that could explain the evolution of clutch size of both high- and low-latitude species.

During the past 50 years, however, the number of ad hoc hypotheses explaining the evolution of clutch size has proliferated (Murray 2001). The current explanation is a series of ad hoc hypotheses, "The explanation of clutch size is (hypothesis A) food limitation, OR (hypothesis B) predation, OR (hypothesis C) something else." These ad hoc hypotheses explaining the evolution of clutch size cannot be tested, that is, they cannot be refuted. If a biologist studies the clutch size in species X, he may conclude that his evidence "supports hypothesis A" or, if it does not, it "supports hypothesis B," or, if it does not, it "supports hypothesis C." Biologists do not conclude that their data on hypothesis B refute hypothesis A, even when it does. They do not say, "We biologists should reject hypothesis A because the data support hypothesis B." For example, biologists do not usually conclude that their data, based on their study of coots, should refute a hypothesis based on a study of waterfowl, even when they do, because there is no reason for them to believe that one hypothesis should be able to explain clutch size in both coots and waterfowl. A biologist is either a unifier, or he is not, and this is a personal choice.

Much ink has been consumed in discussing evolution's most famous "tautology": the phrase, "survival of the fittest." Clearly, however, this is not a tautology because we have to know the meanings of the words to make sense of it (perhaps the reason why Popper [1972] called it "almost tautological"). "Survival of the fittest" is a slogan, which was intended to capture the essence of natural selection in a short phrase. The real problem with "survival of the fittest," as a description of the evolutionary process, is its vagueness. The phrase, "survival of the fittest," predicts nothing that we can observe. Even if what we observe is the fittest, How do we recognize it as the "fittest?" Furthermore, are we referring to the fittest genotype, the fittest individuals, the fittest group (e.g., family ["kin selection"]), or the fittest population? Are we discussing the survival of individuals, groups, or genotypes?

Determinism

In physics this term has been applied to the notion that a theory provides explicit predictions that may or may not be corroborated by empirical facts. In particular, Newton's theory of motion is deterministic because, given the laws and certain initial conditions, a prediction is that a body near the earth's surface will fall 16.1 feet in the first second of free-fall (all of the time, not 50%, 75%, or 95% of the time). Material bodies either fall 16.1 feet, or they do not. If they do not, Newton's laws are falsified. If they do, and they did ("not yet falsified"), the theory gains support. Newton's laws are not proved by an experiment showing that material bodies fall 16.1 feet in the first second of free-fall near the earth's surface; they just have not yet been falsified. (Newton was aware that such things as feathers did not fall 16.1 feet in the first second of free-fall, and he was aware that a bit of fluffy down and a piece of gold descended with the same velocity in vacuums experimentally developed by Boyle. The fact that feathers fall more slowly in the earth's atmosphere than a piece of gold does not refute Newton's theory.)

The success of Newton's theory led to a second notion of determinism, usually attributed to Laplace—that if you knew the initial positions and velocities of all the particles in the universe, you could predict the future behavior of the universe (Rothman 1995), including the thoughts, feelings, and behavior of human beings (Popper 1979; Rose et al. 1984). This broader interpretation of determinism implies predetermination and an absence of free will. Once the universe was set in motion its history rolled out as a movie, each frame of which was strictly and logically determined by the first frame (Popper 1982). Human behavior today is supposed to be determined by the positions and velocities of particles millions of years ago. This notion of determinism, including its more invidious forms—social Darwinism and biological determinism—has generally been rejected by biologists.

One should be absolutely clear that this second notion of determinism is purely philosophical—the rather extraordinary extrapolation of the determinism *perceived* in physical theory and experiments to the biological world. There is not and never was the slightest empirical support for extending (by induction) the determinism of particular theories applied to particular physical situations to the determinism of human behavior. It should be noted that, despite being the most outspoken promoter of the development of deductive science, Popper (1979, 1982) forcefully rejects the notion of determinism as a substantial basis for science, much less as a basis on which to rationalize injustices exhibited in human behavior. He states that he himself is an indeterminist and proposed that even Newton was an indeterminist.

Popper is probably correct regarding Newton's being an indeterminist. If the hypothesis is that one could predict the future state of the universe, much less human behavior, from knowledge of the present positions and velocities of all particles, Newton would probably reject it, given his distain for hypotheses that were without empirical support. His famous quote, "Hypotheses non fingo" ("I make no hypotheses") referred specifically to the problem of *why* physical bodies were attracted to one another. If he could not bring himself to speculate about the cause of the gravitational force, it seems unlikely that he would have entertained any hypothesis connecting human behavior to physical laws.

What we should not do is confuse the first meaning of determinism with the second. We should be able to propose laws and deduce their consequences in specific instances (i.e., propose deterministic theory) without claiming further that if we knew even more we could predict the future completely. The latter hypothesis is an unjustified extension of the former. Thus, we are not justified in rejecting deterministic theory solely on the basis of our rejecting determinism in its second sense.

When it comes to ideas, confusion is unbounded. Johnstone (1986) claims that Popper holds a "deterministic interpretation of cause and effect," and, because I cited Popper favorably in one of my papers (Murray 1986), he attributes the same philosophy to me. But, in his essay *On Clouds and Clocks*, Popper (1979) argued forcefully that physical determinism was a flawed philosophy and twice declared himself to be an "indeterminist." Indeed, Popper (1982) has written a whole book on indeterminism; its subtitle, *An Argument for Indeterminism*. You may disagree with Popper's arguments, but you certainly cannot characterize him as a "determinist." What can Johnstone have in mind, then, when he labels both Popper and me 'determinists'?" Has he used one definition of the word, and we another? How can we scientists make progress in understanding nature if we (even unwittingly) misread, misinterpret, or misrepresent the ideas of others, especially those with whom we disagree (see Chapter 7)?

The world is not deterministic. What are deterministic are theories that explain an indeterministic world. Theories must be deterministic; otherwise they could not be tested.

Statistics

Statistics is a method for determining whether the result of some observations or experiments is "true" or not. One makes a Type I error when a true hypothesis is rejected, or a Type II error when an actual difference is missed.

Statistical tests in ecology and evolution are overused. What gets lost among the statistics are the biological results. In many papers, we are treated to the results of statistical tests but not the data. In ecology and evolutionary biology especially, where we are often working with small samples, I am sure that Type I and Type II errors are frequent. What we should always keep in mind is that statistical tests do not pave the road to scientific truth.

In 1989 I published two papers in *Evolution* on the evolution of clutch size (Murray and Nolan 1989, Murray et al. 1989). In these papers, my coauthors and I predicted, on the basis of other demographic parameters, that the mean clutch size of the Prairie Warbler *Dendroica discolor* was 3.49 (empirical, 3.89) and that of the Florida Scrub-Jay *Aphelocoma coeruleus* was 3.43 (empirical, 3.33). As far as I know, no one had ever before made a quantitative prediction of a biological fact from theory. I was instructed by the editor that I would have to write a "sensitivity analysis," whatever that was (or is), before my papers would be accepted. I wrote some claptrap as an appendix, and my papers were published.

In my theoretical discussions I can predict a number, say mean clutch size, to any number of decimal places I choose. No statistics are necessary. As I recall, the "sensitivity analysis" (or at least what I wrote) was a discussion of how variations in the input numbers would affect the predicted number. This was a whole lot of nonsense that I could or would not deal with—so I wrote claptrap. Statistics should be calculated from observational or experimental data in order to determine the standard deviation or standard error of the sample. The question is, or should be, Does the predicted value lie outside two standard errors? Anyone who thought that the predicted value of 3.43 was outside two standard errors of a sample mean of 3.33 does not know much biology or, even, statistics. Curiously, despite the success of the Murray-Nolan clutch-size equation in predicting a quantitative value for the mean clutch size of three species and explains several conundra of clutch-size variations, the equation has been of no interest to evolutionary biologists.

Criticism

According to Karl Popper (1989:vii),

Criticism of our conjectures is of decisive importance: by bringing out our mistakes it makes us understand the difficulties of the problem which we are trying to solve. This is how we become better acquainted with our problem, and able to propose more mature solutions: the very refutation of a theory—that is, of any serious tentative solution to our problem—is always a step forward that takes us nearer to the truth. And this is how we can learn from our mistakes.

EEBs are humans, and no human enjoys having his ideas criticized. Nevertheless, EEBs, in general, *detest* criticism. They engage in criticism only as anonymous reviewers of manuscripts submitted for publication. They, of course, under the protection of their editors, do not reply to criticism by authors of the prepublication reviews. The whole issue of anonymity is intended to avoid accountability by reviewers and editors who hide behind each other (in peer review, but that is another story).

First impressions

We all know the importance of first impressions. This is especially true when we are interviewing for a job that we want. Or meeting a person whom we want to see again. First impressions are important in science as well. Elias (1971), an anatomist, pointed out how first impressions influenced the development of subsequent interpretations of microscopic sections. The livers of vertebrates had been described as three-dimensional networks of crooked, interconnected cylinders for over 100 years, despite the fact that no microscopist had ever found a cross section or an oblique section. Elias suggested, "Those who do not believe in fairies and genies [that arranged microtomes and cylindars always to obtain longitudinal sections] must find a different explanation for the hepatic miracle." It turns out that the liver is constructed from interconnecting sheets, much like the walls of a building. A scientist can at least wonder how many other first impressions control our thinking.

An ecological example comes from the first serious experimental study of population dynamics (Gause 1934), the growth of *Paramecium* populations from near zero size to near maximum size was described as logistic, characterized by a nice *S*-shaped curve, which was superimposed over the data, as shown in virtually every ecological textbook. Gause was influenced by Pearl's (1925) proposal of logistic growth being a law of population dynamics.

Suppose that Gause, instead of *assuming* logistic growth and "fitting" a logistic equation, had instead simply plotted a quadratic regression to his data on *Paramecium aurelia* from its near zero to its maximum population size. He would have obtained the following result ($R^2 = 0.9877$, $p = 0.032$).

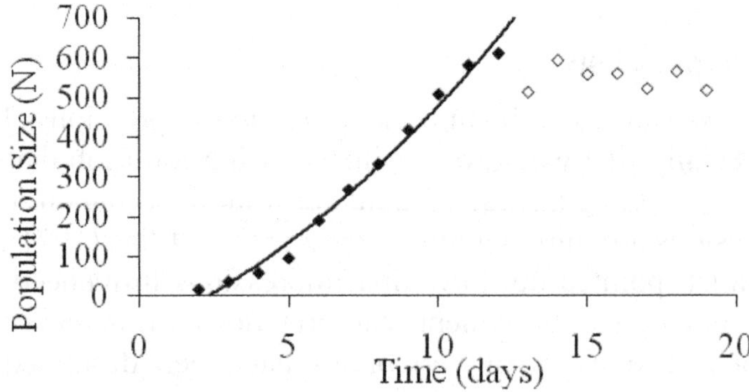

Do you suppose that any subsequent ecologist would have thought that logistic growth (inflection point at $N = 305$) were a possibility?

We shall take up logistic growth in the next chapter. For the moment, we should wonder how much of what we "know" is a consequence of careful research or the passive acceptance of first impressions.

Innumeracy

Americans are innumerate (Paulos 1988). Although many can add, subtract, multiply, and divide, and may even fill out their own income tax forms, the real trick in mathematics is to know *when* to add, subtract, multiply, and divide *what* numbers. We often hear that the divorce rate in the United States is about fifty percent. Where does this number come from? I got a clue from Nora Ephron (2004), who wrote, "What is true is that every year there are twice as many marriages as divorces..." Although this statement may be true, it does not mean that the divorce rate is 50% per year. In order to determine the annual divorce rate, we must divide the number of divorces during year X by the number of married couples at the beginning of year X. This comes to a divorce rate of about two percent per year. At two percent per year, the "half-life" of a marriage is 35 years. Thus, fifty percent of the married couples counted at the beginning of year X will be married to each other for 35 years, unless one or the other spouse dies. Some will be married for a shorter period, others for a longer period.

Scientists do not do much better. For example, one datum that ornithologists want to know is the percentage of clutches that are expected to be successful in producing young that leave the nest. That seems simple enough: count the number of clutches that you have found, count the number of clutches that produce young to leave the nest, and divide the latter by the former.

Mayfield (1960, 1961, 1975), however, was concerned that ornithologists were overestimating nest success, that is the probability that any young will leave a nest successfully. Ornithologists find nests in various stages of the nest episode, from nest building through egg laying, incubation of eggs, and brooding of nestlings to the fledging of young. A higher proportion of nests found late in the nesting episode should be successful than of nests found early in the nesting episode: a smaller percentage of nests found early in a nest episode would successfully produce young than the same number of nests found just before young usually leave the nest. Instead of dividing the number of successful clutches by the number of clutches found, Mayfield suggested that ornithologists calculate the mortality rate of nests by dividing the number of failed nests by the number of "exposure days" (a nest observed for five days contributes five "exposure days"; one observed for 10 days contributes 10 "exposure days" to the total). The nest survival rate is obtained by subtracting the mortality rate from one. He obtains the clutch success rate by raising the survival rate to the power "length of nest period," that is,

$$percentage\ of\ clutches\ that\ are\ successful = (1 - mortality\ rate)^{length\ of\ nest\ period}.$$

(1.1)

The Mayfield method is a bit more complicated than this. For example, he calculates success during the incubation stage and brooding stage separately and, then, multiplies the two to obtain "nest success." Mayfield's papers explaining the method of calculating the success of clutches in producing hatchlings, producing nestlings, and producing fledglings is probably the most cited ornithological paper in history. There are 703 citations since 1993 (until the time of my writing, April 2006; 1271 for the 1975 paper, alone, to Sept. 2010, JRJ). Despite elaborate statistical analyses, no one seems to have caught what is a mistake in high school algebra.

The problem with the Mayfield method is that it is wrong. When he divides the number of failed nests by the number of exposure days, he obtains the *number* of failed nests per exposure day, not the *percentage* of failed nests per exposure day. Thus, "1 − mortality rate" in Eq. 1.1 is the *number* of nests that survive per exposure day, not the *percentage* of nests that survive per exposure day. In statistics, the probability of an event (e.g., the success of a group of clutches over a period of time—the nest episode) is the product of the probabilities of each event (day to day probability of success), as correctly expressed in Eq. 1.1. The Mayfield method uses the wrong estimate of daily survival probability.

I do not fault Mayfield for his errors. Everyone makes mistakes. I fault the hundreds, if not more than a thousand ornithologists/ecologists/evolutionary biologists, who have followed the Mayfield method over the past 45 years without checking the mathematics.

Even sophisticated mathematicians make mistakes. Statistical demographer Lotka (1925) defined a population's per capita birth rate (b) as (in my notation),

$$1/b = \int_0^\infty l_x e^{-rx}\, dx, \qquad\qquad (1.2)$$

where l_x is the probability of surviving from birth to age x, r is the per capita growth rate, and e is the base of the natural logarithms. This is also wrong, but I will not explain why here (see Chapter 2). I am unaware that the per capita birth rate has been determined with this equation for any population. Mathematical ecologists discuss "birth rate" often, but seem uninterested in actually calculating a numerical value for the birth rate of the populations they are studying. (I may be wrong in this. The literature is enormous. A calculation may have been done but is buried somewhere.) In this case, a skilled mathematician did not check the value of the birth rate calculated with this equation with the numerical value determined with a second method of calculating birth rate (we shall do this in Chapter 3). This is a common error among biomathematicians today, at least by those working in the areas of theoretical population dynamics I am (ecology and evolution). They create mathematical equations or models that seem to bear no relationship to empirical data. This will be a major theme throughout this book.

Counterexamples

Theoretical (i.e., non-applied) mathematicians spend considerable time proving theorems by deducing them from axioms. Theorems that have not been deduced from axioms are essentially ad hoc relationships. For example, consider the following problem. Mark two points on the circumference of a circle and join them with a line. How many spaces are formed within the circle's circumference? Pretty easy; there are two spaces. Now, mark three points and join them with straight lines. How many spaces are formed? Four. Now, mark four points and join them with straight lines. How many spaces are formed? Eight. As we add points to the circumference of the circle, then, we have increased the number of spaces:

Points   2 3 4 5 6

Spaces   2 4 8 ? ?

We have a pattern emerging. With each additional point, we double the number of spaces within the circle. May we pose this relationship as an axiom? Of course, any statement can be proposed as an axiom. What we must do, however, is test the axiom by making a prediction. The prediction from this axiom is that with five points joined with straight lines, there should be 16 spaces within the circle. We count the number of spaces: sixteen. Look at what we have done. We have collected data, we have generated a hypothesis (the axiom) from which we made a prediction (16 spaces), and we have tested both by comparing the prediction with observation. Our axiom has accurately predicted the number of spaces within the circle. We have done all the things one is supposed to do to develop a new theorem. But, how much confidence should we have in our hypothesis (axiom)? In this case, not much because when six points are joined by lines, the circle is divided into 30 spaces, not 32. Our hypothesis is a failure as a generalization and illustrates the problem with all inductive hypotheses. We can never know them to be universally "true."

We should consider ourselves lucky. We discovered our error early on in our research on this problem. But, how many cherished biological generalizations, established by induction from a limited amount of empirical data (or, worse, established by an erroneous interpretation or based on faulty data), remain today unchallenged by investigators?

A famous theorem in mathematics is Fermat's Last Theorem. It states that there are no whole number solutions for the equation, $x^n + y^n = z^n$ when $n > 2$. Fermat claimed to have proved it but left no written proof. Many mathematicians (and nonmathematicians) have attempted proofs but failed. Even though mathematicians had shown that Fermat's Last Theorem was true for values of n up to 4,000,000 (Singh 1997), mathematicians could not say that Fermat's Last Theorem was true for all values of n. Maybe n = 4,000,001 or 5,000,000 would provide the counterexample disproving Fermat's Last Theorem. Andrew Wiles, however, committed seven years of his life to deducing Fermat's Last Theorem from mathematical axioms and eventually provided a long and difficult proof (Singh 1997). We now know that Fermat's Last Theorem is true because it follow's logically from the accepted theorems of mathematics. &&&&

As Dunham (1994:121, bold italics in original) pointed out, "To prove a general statement [in mathematics] requires a general argument; to disprove it requires but a single instance in which the statement fails. The latter is called a ***counterexample***, and a good counterexample is worth its weight in gold." Biology is rife with counterexamples to the best of its general statements. Almost all of the 100 inductive generalizations identified by Rensch (1971) have "exceptions," which are, of course, the counterexamples refuting the generalization (whether rule, principle, or law) as being universally true.

Biologists, even those with a mathematical background, seem not to practice the method of designing counterexamples to evaluate their beliefs. I think this is because they are happy with untestable ad hoc hypotheses (see Dyson, 1988: 45-46). Their logic runs from "axioms" to "theorems" (i.e., predictions), but there is no desire for them to see whether their predictions describe any real biological phenomena. It is enough to deduce theorems from axioms, whether or not even the axioms correspond to biological reality.

I cannot say that we should be as cynical as Andreski (1972: 129-130) is about mathematics in the social sciences,

Most of the applications of mathematics to the social sciences outside economics are in the nature of ritual invocations which have created their own brand of magician. The recipe for authorship in this line of business is as simple as it is rewarding: just get hold of a textbook of mathematics, copy the less complicated parts, put in some references to the literature in one or two branches of the social studies without worrying unduly about whether the formulae which you wrote down have any bearing on the real human actions, and give your product a good-sounding title, which suggests that you have found a key to an exact science of collective behaviour.

I think I can say with confidence that the mathematics in population biology comes close.

Inasmuch as my ideas and equations have not been cited, much less discussed in the biological literature, I had always suggested that my colleagues "cannot do the arithmetic." When I am more mellow, I would say they were willfully not doing the math. I have come to believe that the problem with my colleagues is conceptual. They are certainly able to do the high school algebra required to understand my equations. They can multiply (or divide, add, or subtract) two numbers. The conceptual problem is understanding *whether* one is supposed to add, subtract, multiply, or divide the two numbers.

Humility

In *The Pleasure of Finding Things Out*, Richard Feynman (1999) tells two stories about himself that are of interest. Both occurred while he was working on the atom bomb at Los Alamos in the 1940s. Feynman, who had just finished his Ph.D., describes how his one-on-one meeting with Niels Bohr came about. Bohr had come to Los Alamos for a visit. While there he attended a meeting with the scientists to talk about the problems with the bomb. Feynman was in a corner in the back of the room, where he had a limited view. On Bohr's next visit, Feynman received a phone call from Aage Bohr, Niels's son, "My father and I would like to speak with you." "Me?" "That's right." Feynman (1999: 88, italics in original) wrote,

the last time he [Niels Bohr] was there [Los Alamos] he said to his son [Aage]—"Remember the name of that little fellow in the back over there? He's the only guy who"s not afraid of me, and will say when I've got a crazy idea. So *next* time when we want to discuss ideas, we're not going to be able to do it with these guys who say everything is yes, yes, Dr. Bohr. Get that guy first, we'll talk with him first.

The message of this story is that imaginative scientists who generate one new idea after another know that some of their ideas will be bad ones. They *want* criticism, and they prefer to have it before going public. They seek out other thinkers who are interested in discussing ideas for an honest evaluation.

At another time, Feynman had been working on a difficult problem when Enrico Fermi came to visit Los Alamos. About his meeting with Fermi, Feynman (1999: 85-

86, italics in original) wrote about a problem he was working on,

The calculations were so elaborate it was very difficult. Now, usually, I was the expert in this, I could always tell you what the answer was going to look like or when I got it I could explain why. But this thing was so complicated I couldn't explain *why* it was like that. So I said to Fermi that I was doing this problem and I started to calculate—he said, wait, before you tell me the result, let me think. Its going to come out like this (he was right), and its going to come out like this because of so and so. And there's a perfectly obvious explanation. . . . So *he* was doing what I was supposed to be good at, ten times better. So that was quite a lesson to me.

What was the lesson? Feynman did not say, but my guess is that no matter how good you think you are, there is someone around who is better. Logically, there must be one "best," but if you think you are the best, there is a good chance you may be mistaken. Thus, again, this points to the value of open discussion with those who disagree with you.

Biologists tend to think differently. For example, Lawton (1992) warns ecologists, especially older ecologists, to be aware of the Infallibility Syndrome (IS), the condition of believing that you are always right in your thinking about just about everything. He advises graduate students, "Take as much advice as you can about your work, but take it all with a pinch of salt, particularly when it is offered by somebody with IS." He also tells us this story:

If you think you have a really new idea, pause and think about [sic]; it is probably something published as long ago as 1965, or even more obscurely, in the *Origin of Species*. One example will suffice. At a recent international meeting I attended, one newly graduated Ph.D. told me perfecty seriously that there was very little evidence for density dependent population regulation. Such appalling ignorance is not dissimilar to believing that the world is flat, or that evolution is a myth, both beliefs that are easy to sustain if you don't actually know anything.

Sadly, Lawton shows that there is no humility and lots of IS here. Of course, Lawton notes that those afflicted with IS do not know it. Instead of being dismissive, perhaps Lawton should have responded to this Ph.D. with, "Why do you think that?," to which I would have replied, "Inasmuch as I cannot prove a negative, why don't you give me your best evidence for density-dependent regulation?" Indeed, I have offered reasons for thinking there is no evidence for density-dependent regulation and have offered alternative models (Murray 1979, 1982, 1994, 1999a,b, 2000a,b, 2001). So far, Lawton and no one else has taken the opportunity to show that I am wrong.

With regard to EEBs, I do not know (on the basis of what they have written, whether in print or in letters to me) of anyone (other than my long-time friend, Joe Jehl), including freshly minted Ph.D.s, who are even slightly interested in discussing ideas for the sake of discussing ideas. EEBs are advocates of beliefs rather than evaluators of ideas.

Belief

One dictionary definition of belief is, "confidence in the truth or existence of something not immediately susceptible to rigorous proof" (The Random House Dictionary of the English Language, College Edition, 1968). Scientists usually set aside the idea of belief as referring to religious beliefs. We rarely describe our own unsubstantiated conjectures as "beliefs." We all subscribe to the notion that scientific utterances are based on evidence rather than belief. But consider the following statements with regard to the truth of density-dependent regulation:

(1) "... the absence of field evidence does not, and will not, make the advocates of density-dependent regulation change their minds . . . because, given certain assumptions about the persistence of natural populations, the existence of density-dependent regulation becomes a logical necessity" (Lack 1966:291);

(2) "The concept of density-dependent regulation ... is derived through pure logical deduction from one major premise, namely the persistence of populations in the wide sense. It is knowledge obtained a priori. There is no need to test its validity against observations ..." (Royama 1977:33);

(3) "Royama (1977) has argued convincingly that tests for density-dependence are unnecessary" (Berryman 1991:142); and

(4) "... density dependence is a research program, a theoretical framework with which to investigate the causal factors of population fluctuations, not a hypothesis to be falsified or corroborated (Turchin 1999:157).

What could be clearer statements of belief? of advocacy? The notion that evidence is unnecessary is dogma, not science. DDRHs are guesses without supporting evidence. In fact, Andrewartha and Birch (1954:649) explicitly stated that density-dependent factors,

have a peculiar logical status. They are not a general theory, because ... they do not describe any substantial body of facts. Nor are they usually put forward as a hypothesis to be tested by experiment and discarded if they prove inconsistent with empirical fact. On the contrary, they are usually asserted as if their truth were axiomatic .... These generalizations about "density-dependent factors" and competition in so far as they refer to natural populations are neither theory nor hypothesis but dogma.

Lack (1966:291) replied that Andrewartha and Birch's characterization of density-dependent regulation as dogma was a "misuse" of the term, which has its "proper meaning in [a] different intellectual discipline." But Lack is wrong; a dictionary definition of "dogma" is, "a settled or established opinion, belief, or principle" (The Random House Dictionary of the English Language, College Edition, 1968). What could be more dogmatic than a principle lacking, according to Lack himself, substantiating evidence?

The sad tale of ecology is that the story has not changed in the last 40 years.

I am not arguing that the biologists' inductive generalizations are not science. Every science must start this way, describing and organizing the apparent facts into patterns. Furthermore, most scientists are engaged in what Kuhn (1962) characterizes as normal science. They are solving puzzles that bring experience into line with prevailing theory. Few physicists in the eighteenth and nineteenth centuries were expecting to bring down Newton's theory with their experiments. Revolutions occur only infrequently in science, and few scientists have the opportunity to conduct the crucial experiment or make the crucial observation that swings scientific opinion in favor of a new theory. An individual investigator's failure to solve a particular problem is often taken as his or her personal failure in designing and carrying out the experiment, rather than a failure of the theory (Kuhn, 1970). Nevertheless, each puzzle solved lends further support for a theory.

Biologists, by contrast, are not usually testing the hypothetical consequences that have been logically deduced from the laws of some theory, if only because they recognize no Newtonian-like biological laws. As a result, they simply "test" *ad hoc* hypotheses, which develop out of their experience and which are within the acceptable range of answers to puzzles within the metaphysical research program called natural selection. They are often answering the question, "How do I explain my data in terms of the theory of natural selection?" (Murray1986, 1991).

This state of affairs in biology, the Baconian exploration of nature, is perfectly justifiable as long as the investigators understand that they are not doing deductive science. What I am arguing is that biologists and philosophers of biology eschew a research method that has been so successful in physics—the deductive method so articulately described by Popper (1968, 1979, 1989). As noted above, some biologists excuse their failure to develop such laws and theory by claiming that biological systems are far more complex than physical systems. Others claim that the deductive method of the physicists is inappropriate for evolutionary theory, which is essentially a historical science (Mayr, 1982, 1991, 1996). Furthermore, biologists presume that the deductive method implies determinism, a philosophy they reject (Mayr, 1982, 1988;

Rose, Kamin, and Lewontin, 1984). For them, deduction is inappropriate in biology because the future cannot be predicted.

The issue would seem to be whether biologists should be able to develop a predictive theory with universal laws. To hypothesize that they cannot because they have not (or for any other reason) is an inductive inference in itself. I happen to think that this hypothesis is wrong.

Progress in science occurs in three stages: discovery, justification, and acceptance. There are no rules or methods that guarantee discovery. Through hard work and luck a scientist may discover a new species, structure, relationship, behavior, law, or whatever in nature. The scientist must then justify his or her discovery. This may be rather straightforward in the case of a new species or structure. Justifying new explanations (such as inertia or natural selection, which cannot be seen) is more difficult and involves the rules of logic. Why one scientist accepts a new theory when another does not is not easy to explain, being a psychological phenomenon. Each thinks the other is doing bad science, even when both are exposed to the same observations and arguments. Again, there are no rules.

Literature Cited

Aarssen, L. 1997. On the progress of ecology. Oikos 80;177-178.

Andreski, S. 1972. *Social Sciences as Sorcery*. St. Martin's Press, New York.

Andrewartha, H. G., and L. C. Birch. 1954. *The Distribution and Abundance of Animals*. University of Chicago Press, Chicago, IL.

Ankney, C. D., A. F. Afton, and R. Alisauskas. 1991. The role of nutrition reserves in limiting waterfowl reproduction.Condor 93:1029-1032.

Arnold, T. W., F. C. Rowher., and T. Armstrong. 1987. Egg viability,nest predation, and the adaptive significance of clutch size in prairie ducks. Am. Naturalist 130-643-653.

Bartholomew, G. A. 1982. Scientific innovation and creativity: a zoologist's point of view. American Zoologist 22:227-235.

Bartholomew, G. A. 1986. The role of natural history in contemporary biology. BioScience 36:324-329.

Begon, M. 1998. The vole *Clethrionomys rufocanus* - a modern classic? Researches on Population Ecology 40:145-147.

Berryman, A. A. 1991. Stabilization or regulation: what it all means! Oecologia 86:140-143.

Calaprice, A. (ed.). 2005. *The New Quotable Einstein.* Princeton University Press, Princeton.

Carroll, L. 1960. *Alice's Adventures in Wonderland & Through the Looking-Glass.* New York, Signet.

Christie, A. 1929. *Partners in Crime.* Berkley, New York.

Cody, M. 1966. A general theory of clutch size. Evolution 20:174-184.

Cohen, J. E. 1971. Mathematics as metaphor. Science 172:674-675.

Cohen, I. B. 1980. *The Newtonian Revolution.* Cambridge University Press, Cambridge, England.

Corcoran, J. 1995. Tautology *in* The Cambridge Dictionary of Philosophy (R. Audi, ed.). Cambridge University Press, Cambridge.

Crick, F. 1988. *What Mad Pursuit: A Personal View of Scientific Discovery.* Basic Books, New York.

Dunham, W. 1994. *The Mathematical Universe: An Alphabetical Journey through the Great Proofs, Problems, and Personalities.* Wiley, New York.

Dyson, F. J. 1988. *Infinite in All Directions.* Harper & Row, New York.

Einstein, A. 1954. *Ideas and opinions.* Crown, New York.

Einstein, A., and L. Infeld. 1938. *The evolution of physics: from early concepts to relativity and quanta.* Simon & Schuster, New York.

Egler. F. E. 1986. "Physics envy" in ecology. Bull. Ecol. Soc. Amer. 67:233-235.

Elias, H. 1971. Three-dimensional structure identified from single sections. Science 174:993-1000.

Ephron, N. 2004. Can marriage be saved? A forum. The Nation 279:19-20.

Fagerstrom, T. 1987. On theory, data, and mathematics in ecology. Oikos 50:258-261.

Fahrig, L. 1988. Nature of ecological theories. Ecological Modelling 43:129-132.

Feynman, R. 1965. *The Character of Physical Law.* MIT Press, Cambridge, MA.

Feynman, R. P. 1985. *QED: The Strange Theory of Light and Matter.* Princeton Univ. Press, Princeton, NJ.

Feynman, R. P. 1995. *Six Easy Pieces: Essentials of Physics Explained by Its Most Brilliant Teacher.* Addison-Wesley, Reading, MA.

Feynman, R. P 1999. *The Pleasure of Finding Things Out.* Basic Books, New York.

Gause, G. F. 1934. *The Struggle for Existence.* Hafner, New York. (reprint 1964).

Hempel, C. G. 1965. *Aspects of scientific explanation and other essays in the philosophy of science.* The Free Press.

Hempel, C. G., and P. Oppenheim. 1948. Studies in the logic of explanation. Philosophy of science 15:135-175.

Johnstone, I. M. 1986. Plant invasion windows: a time-based classification of invasion potential. Biological Reviews 61: 369-394.

Kuhn, T. S. 1962. *The structure of scientific revolutions.* Univ. of Chicago Press, Chicago, IL.

Kuhn, T. S. 1970. Logic of discovery or psychology of research. *In* Latkos, I., and A. Musgrave (eds.), *Criticism and the growth of knowledge.* Cambridge Univ. Press, pp. 1-23.

Lack, D. 1947. The significance of clutch-size. Parts I and II. Ibis 89:302-352.

Lack, D. 1948. The significance of clutch size, Pt. III. Ibis 909:25-45.

Lack, D. 1954. *The natural regulation of animal numbers.* Oxford Univ. Press. London.

Lack, D. 1966. *Population Studies of Birds.* Clarendon Press, Oxford.

Lawton, J. H. 1992. (Modest) advice for graduate students. Oikos 65:361-362.

Lawton, J. H. 1999. Are there general laws in ecology. Oikos 84:177-192.

Livio, M. 2002. *The golden ratio.* Broadway Books, New York.

Lotka, A. J. 1925. *Elements of Physical Biology.* Williams & Wilkins, Baltimore.

Mayfield, H. 1960. *The Kirtland's Warbler.* Cranbrook Institute of Science, Bloomfield Hills, MI.

Mayfield, H. 1961. Nesting success calculated from exposure. The Wilson Bulletin 73:255-261.

Mayfield, H. F. 1975. Suggestions for calculating nest success. The Wilson Bulletin 87:456-466.

Mayr. E. 1982. *The Growth of Biological Thought: Diversity, Evolution, and Inheritance.* Harvard Univ. Press, Cambridge, MA.

Mayr, E. 1988. *Toward a New Philosophy of Biology.* Harvard University Press, Cambridge, Massachusetts.

Mayr, E 1996. The autonomy of biology: the position of biology among the sciences. Quarterly Review of Biology 71:97-106.

Murray, B. G., Jr. 1979. *Population Dynamics: Alternative Models.* Academic Press, New York.

Murray, B. G., Jr 1982. On the meaning of density dependence. Oecologia 53:370-373.

Murray, B. G., Jr 1986. The structure of theory, and the role of competition in community dynamics. Oikos 46:145-158.

Murray, B. G., Jr 1991 . Sir Isaac Newton and the evolution of clutch size: a defense of the hypothetico-deductive method in ecology and evolutionary

biology. Pp. 143-180 *in Beyond Belief:Randomness, Prediction, and Explanation in Science* (J. Casti and A. Karlqvist, eds.). CERC Press, Boca Raton, FL.

Murray, B. G., Jr 1994. On density dependence. Oikos 69:520-523.

Murray, B. G., Jr 1999a. Can the population regulation controversy be buried and forgotten? Oikos 84:148-152.

Murray, B. G., Jr 1999b. Is theoretical ecology a science? A reply to Turchin (1999). Oikos 87:594-600.

Murray, B. G., Jr 2000a. Dynamics of an age-structured population drawn from a random numbers table. Austral Ecology 25:297-304.

Murray, B. G., Jr 2000b. Density dependence: reply to Tyre and Tenhumberg. Austral Ecology 25:308-310.

Murray, B. G., Jr 2001. Are ecological and evolutionary theories scientific? Biological Reviews 76:255-289.

Murray, B. G., Jr., and V. Nolan, Jr. 1989. The evolution of clutch size. I. An equation for predicting clutch size. Evolution 43:1699-1705.

Murray, B. G., Jr., J. W. Fitzpatrick, and G. E. Woolfenden. 1989. The evolution of clutch size. II. A test of the Murray-Nolan equation. Evolution 43:1706-1711.

Northrop, F.S.C. 1947. The logic of the sciences and the humanities. Meridian Books, Inc. NY, NY.

Paulos, J. A. 1988. *Innumeracy: Mathematical Illiteracy and its Consequences.* Hill and Wang, New York.

Pearl, R. 1925. *The Biology of Population Growth.* Alfred A. Knopf, New York.

Popper, K. R. 1959. *The Logic of Scientific Discovery.* Basic Books, New York.

Popper, K. R. 1968. *The Logic of scientific discover.* Harper & Row, New York.

Popper, K. R. 1972. *Objective Knowledge.* Clarendon Press, Oxford.

Popper, K. R. 1979. *Objective Knowledge: An Evolutionary Approach* (revised ed.). Oxford University Press, London.

Popper, K. R. 1982. *The Open Universe: An Argument for Indeterminism.* Routledge, London.

Popper, K. R. 1989. *Conjectures and Refutations: the Growth of Scientific Knowledge.* Routledge, London.

Quenette, P. Y., and J. F. Gerard. 1993. Why biologists do not think like Newtonian physicists. Oikos 68:361-363.

Quinn, J. F., and A. E. Dunham. 1983. On hypothesis testing in ecology and evolution. American Naturalist 122:602-617.

Rensch, B. 1971. *Biophilosophy.* Columbia University Press, New York.

Rogers, E. M. 1960. *Physics for the Inquiring Mind.* Princeton Univ. Press, Princeton, NJ.

Rose, S., L. C. Kamin, and R. C. Lewontin. 1984. *Not in Our Genes: Biology, Ideology, and Human Nature.* Harmondsworth, England.

Rothman, T. 1995. *Instant Physics.* Fawcett Columbine, New York.

Royama, T. 1977. Population persistence and density dependence. Ecological Monographs 47:1-35.

Singh, S. 1997. *Fermat's Enigma: The Epic Quest to Solve the World's Greatest Mathematical Problem.* Doubleday, New York.

Skutch, A. F. 1949. Do tropical birds rear as many young as they can nourish? Ibis 91:430-455.

Slobodkin, L. B. 1988. Intellectual problems of applied ecology. BioScience 38:337-342.

Stebbins, G. L. 1977. In defense of evolution: tautology or theory? American Naturalist 111:386-390.

Strong, D. R., Jr. 1980. Null hypotheses in ecology.Synthese 43:271-285.

Strong, D. R., Jr. 1983. Natural variability and the manifold mechanisms of ecological communities. American Naturalist. 122:636-660.

Turchin, P. 1999. Population regulation: a synthetic view. Oikos 84:153-159.

VanValen, L., and F. A. Pitelka. 1974. Commentary—Intellectual censorship in ecology. Ecology 55:925-926.

Chapter 2

Population Dynamics without Regulation: A New Equation

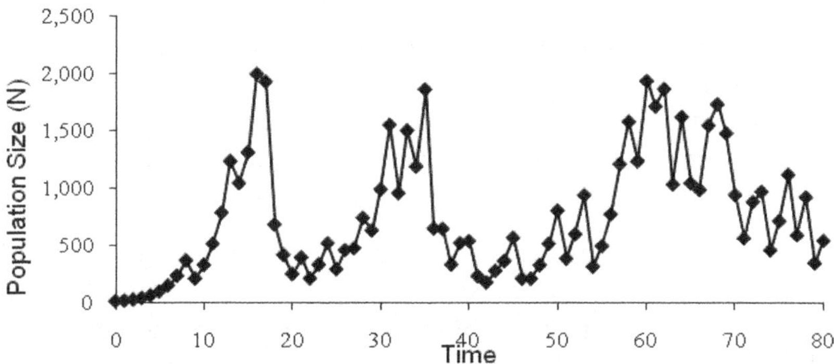

[Despite the lack of truly supportive evidence and an abundance of criticism, the nearly universal acceptance of "density dependence" and the success of the logistic equation as an endlessly manipulatable equation have effectively blinded ecologists from considering alternative models of population dynamics. As evidence, consider the following:

(1) "… the absence of field evidence does not, and will not, make the advocates of density-dependent regulation change their minds … because, given certain assumptions about the persistence of natural populations, the existence of density-dependent regulation becomes a logical necessity" (Lack 1966).

(2) "The concept of density-dependent regulation … is derived through pure logical deduction from one major premise, namely the persistence of populations in the wide

sense. It is knowledge obtained a priori. There is no need to test its validity against observations ..." (Royama 1977).

(3) "Royama (1977) has argued convincingly that tests for density-dependence are unnecessary" (Berryman 1991).

Writing about ecological theory in general, Fagerström (1987) suggested that "as a matter of historical record—it is standard scientific practice to reject ecological data that are in conflict with established theories ... many contemporary ecological theories are in fact retained—and rightly so—although they are demonstrably wrong ... it is by no means obvious that the proximate goal of theorizing should be to find the truth, hence an important role of mathematics is to provide productive lies." (I am aghast! What are scientifically "productive" lies?)

For these four prominent ecologists, the lack of evidence for density-dependent regulation is not sufficient reason for considering alternatives. These statements are barely distinguishable from that of religious Fundamentalists (quoting one, trained as a geologist, from Dawkins 2006), "if all the evidence in the universe turns against creationism, I would be the first to admit it, but I would still be a creationist because that is what the Word of God seems to indicate."

In this regard, would you agree that the population shown in the above graph exhibits density-dependent regulation? It is persistent and fluctuates between an upper and a lower boundary, which are the usual criteria for believing that a population is "regulated." If you are an ecologist you must say "yes," at least if you follow the scientific philosophy of Lack, Royama, Berryman, and Fagerström. Alternatively, you may declare that the ideas of Lack, Royama, and Berryman, regarding density-dependent regulation, are old-fashioned and that the modern theory states that a population's size is regulated in some way by both density-dependent and density-independent processes. Furthermore, density-dependence need only occur at some time or place (Sale and Tolimieri 2000). This, of course, is an untestable hypothesis, as Sale and Tolimieri pointed out. How do you test a hypothesis that states that population size may sometimes be a function of density-dependent processes and at other times a function of density-independent processes?

In fact, the above graph was generated by a new population equation, $N_{t+1} = N_t e^{r_{max} - \varepsilon}$, ($\varepsilon \geq 0$), in which there is no feedback loop between any of the parameters and no predetermined maximum or "equilibrium" size, such as the K of the logistic, and no lag times. Population size in the graph is *not* regulated by density-dependent regulating processes.

I have not written this up as a paper because—to tell the truth—I know of no one who is interested, much less interested enough to recommend that it be published. The editors of *Oikos, Oecologia, Theoretical Population Biology, Annales Zoologici Fennici, Austral Ecology,* and *Population Ecology,* and other prominent ecologists have been apprised of this equation and the diversity of its predictions when applied to populations of animals and environments with particular characteristics (see below). Evidently, there was no interest among them because no one even acknowledged receipt. If you are willing to think outside the box, however, you may find the following notes of interest.]

MANUSCRIPT

Population Dynamics without Regulation: A New Equation

In my 1979 book I proposed a new model of population dynamics. When I think about problems, I first visualize them. With regard to population dynamics, I first thought of a group of 10 mice in a barn housing 100 tons of grain. I could not imagine that these mice were suffering from a shortage of food. Each mouse, I guessed, had a surfeit of food. They should be surviving and reproducing about as well as possible. Thus, the growth rate (r) would be at a maximum for the population. Then, I thought about 20 mice with 100 tons of grain. They, too, should be surviving and reproducing about as well as possible. Thus, the growth rate (r) would be at a maximum for the population, which would be the same as when the population numbered 10. Then, I thought about 30 mice with 100 tons of grain. They, too, should be surviving and reproducing about as well as possible. Thus, the growth rate (r) would be at a maximum (r_{max}) for the population, which would be the same as when the population numbered 10 and 20. Sooner or later, of course, the population would become so large that not every individual could survive and reproduce as well as at lower densities. At this point—what I call the **Upper Critical Density** (UCD)—the birth rate (b) begins to decrease, or the death rate (d) begins to increase, or both, resulting in a decrease in the growth rate (r). In some species, there may be a **Lower Critical Density** (LDC). I can imagine that 5, 10 or so mice in 100 tons of grain, might be so spread out that they would have trouble finding one another, and thus not reproduce maximally. Below the LCD, the birth rate may be lower or death rate may be higher than above the LCD (often called the Allee Effect). What I had to do next was translate these thoughts into a model (Figure 1). I call this the **population limitation hypothesis** (PLH) in contrast to the **density-dependent regulation hypothesis** (DDRH).

In this model, given a particular density, we know the birth, death, and, therefore, growth rates of the population. It should be understood that Figure 1 represents an idealized version of an infinite number of combinations of birth and death rates plotted against population size. This is much like Newton's drawing of an ellipse as an idealized version of an infinite number of elliptical orbits of planets and comets.

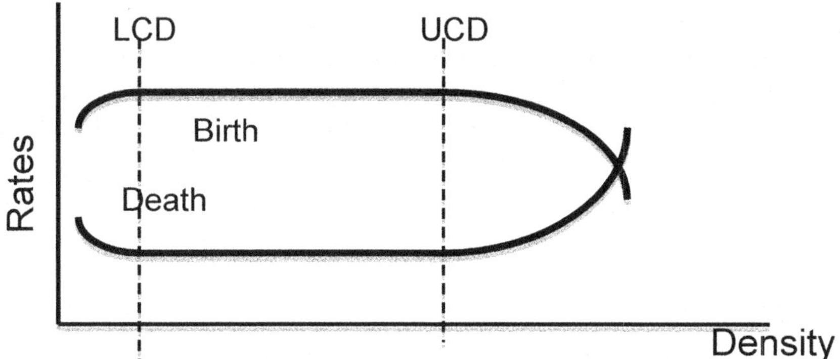

Figure 1. Model of the population limitation hypothesis. LCD is the lower critical density. UCD is the upper critical density.

Newton's drawings did not represent the reality of any particular orbit, and Figure 1 does not represent the reality of any particular population. (I must write this explanation because someone actually submitted a paper to a journal proving that the UCD of my model did not always have to be at 2/3 the maximum population size).

Population Growth Equation

For illustrating how a population could be limited in size and persist in time without the influence of density-dependent regulating factors, let us look at a specific hypothetical but quantitative case. A standard equation for calculating the growth of a population from time t to time $t + 1$ is,

$$N_{t+1} = N_t e^r, \qquad (1)$$

where N_t is population size at time t, N_{t+1} is population size at time $t + 1$, r is the per capita rate of change in numbers, and e is the base of the natural logarithms. A more

general equation is,

$$N_{t+1} = N_t e^{r_{max} - \varepsilon}, \qquad \varepsilon \geq 0 \qquad (2)$$

where r_{max} is the maximum rate of change of a population, which is determined by the evolved individual properties for acquiring, processing, and utilizing resources, and ε is the impact of a reduction of resources on r_{max}. The point at which population growth shifts from Eq. 1 to Eq. 2 is what I have called the upper critical density (UCD; see Figure 1), the point above which the growth rate decreases (Murray 1979, 1982, 1986, 1994).

Food as a limiting factor

Let us first consider a food-limited population. Let us call it Population A. For illustrating how a population could be limited in size and persist in time without the influence of density-dependent regulating factors, let us look at a specific hypothetical but quantitative case.

Now, assume that, as a consequence of the hypothetical species' evolutionary history, individuals consume on average no more than 70 grams of food per capita per day (g/c/d), even when surrounded by an abundance of food. Let us say further that, with an average consumption of 70 g/d, an average individual's probabilities of survival and reproduction are at their maximum values. At this level of consumption, then, the population's per capita growth rate is at its maximum (r_{max}). With less food available per capita, the probabilities of individual survival and reproduction decrease, making r less than maximum. Finally, assume that, with fewer than 50 grams per capita per day (g/c/d), individual survival and reproduction cannot sustain population growth. The population's r becomes negative and its size declines.

The difference between 50 g/c/d, when $r = 0$, and 70 g/c/d, when r is maximum, is 20 g/c/d and represents the maximum amount of resource that the individuals have to devote to population growth. The per capita growth rate of the population between r_{max} and $r = 0$ is a function of the proportional amount of the 20 grams available. The population's growth rate decreases proportionately as per capita consumption decreases.

Quantitative example.—Let us assume that with an average consumption of 70 g/c/d, the growth rate of the population is at its maximum, r_{max}. Say further that $r_{max} = 0.05$. At all densities when food is abundant and average consumption is 70 g/c/d, the population grows exponentially at rate r_{max} (Eq. 1), $N_{t+1} = N_t e^{r_{max}}$. Because the population's growth rate decreases proportionately as per capita consumption decreases, when average per capita consumption is less than 70 g/d, population growth is, $N_{t+1} = N_t e^{r_{max} - \varepsilon}$, $(\varepsilon \geq 0)$.

The difference in mean per capita food consumption between the maximum growth (70 g/c/d) and zero growth (50 g/c/d) is 20 g/c/d. Below the UCD the entire 20 grams is available for population growth. Above the UCD, only a fraction of this 20 g is available for population growth. Thus,

$$\varepsilon = \frac{a-b}{c}, \tag{3}$$

where a is the maximum potential per capita consumption, b is the actual per capita consumption, and c is the range of the amount of food per capita available for growth.

For example, when population size is >UCD and consumption is 65 g/c/d, the reduction of the 20 g is (70 − 65)/20 or 25 percent. If we assume that this results in a corresponding decrease in r_{max}, then $\varepsilon = 0.25 r_{max} = 0.0125$. Thus, from Eq. 2, the population's growth rate between t and $t + 1$ is 0.05 − 0.0125 = 0.0375. When consumption is less than 50 g/c/d, say, 45 g/c/d, availability has been reduced by 125 percent of that required for maximum growth. Thus, $\varepsilon = (70 − 45)/20 \times r_{max} = 0.063$, and $r = -0.013$ (Table 1).

Table 1. Population growth rate (r) for Population A in environments as a function of food availability (in mean grams per capita per day during a year). The population's r_{max} at 70 g/c/d is 0.05. The maximum sustainable size ($r = 0$) occurs when average consumption is 50 g/c/d. The proportional difference is the difference between actual consumption and maximum consumption (i.e., 70 g/c/d) divided by 20. $\square$ = proportional difference times r_{max}.

Per capita food consumption	Actual difference	Proportional difference	$\square$	r
70	0	0.00	0.000	0.050
65	5	0.25	0.013	0.038
60	10	0.50	0.025	0.025
55	15	0.75	0.038	0.013
50	20	1.00	0.050	0.000
45	25	1.25	0.063	0.013
40	30	1.50	0.075	0.025
35	35	1.75	0.088	0.038
30	40	2.00	0.100	0.050

In this model of population dynamics, all we know is the effect that per capita availability of resources has on a population's growth rate. The focus is on the individual's capacity to survive and reproduce, which is a product of its population's evolutionary history, rather than on the capacity of the environment to support it. There is no concept of the environment's carrying capacity (signified K in logistic growth). Furthermore, in order to calculate next year's population size, we do not have to know N_{max}, the population's theoretical maximum or equilibrium size.

An illustration should clarify the dynamics. Suppose that a population is living in a habitat that supplies an average of 100,000 g of resource per day during the year. Accordingly, the N_{UCD} is $100,000/70 = 1,429$. In Table 2, I have shown the transition from densities $< N_{UCD}$ to $> N_{UCD}$. At time 99, for example, $N = 1,412$ and food available per capita is 70.83, that is > 70; thus, $r_{99} = r_{max} = 0.05$. At time 100, $N = 1,484$ and food consumption per capita is 67.38, that is, < 70. This represents a 13.1 percent reduction of the amount of food that can be used for population growth—$(70 - 67.83)/20$. Therefore, $\varepsilon = 0.05 \times 0.131 = 0.0066$, and $r_{100} = 0.0434$. At time 101, $N = 1,550$ and per capita food consumption is 64.51. This represents a 27.43 percent reduction in the amount of food that can be used for population growth—$(70 - 64.51)/20$. Therefore, $\varepsilon = 0.05 \times 0.2743 = 0.0137$, and $r_{101} = 0.0363$. I have plotted the growth of this population from time 0 ($N = 10$) through time 160 ($N = 2,000$) in Figure 2.

Note that the growth curve is *S*-shaped, but it is not logistic. The inflection point, which is at the UCD, is high, about $0.74N_{max}$ in this example—compared with that of the logistic, $0.5K$. Conceptually, N_{max} and K are quite different. N_{max} is the potential maximum population size allowed by the *prevailing* environment—it varies in time and circumstances. It is a consequence of the traits of the organisms and the abundance of the resource. An investigator does not need to know N_{max} before forecasting population growth from N_t to N_{t+1} because N_{max} is not a parameter of either Eq. 1 or 2. In contrast, in logistic growth, K is the hypothetical "carrying capacity of the environment" or the "equilibrium population size," toward which the population tends to grow, increasing when below K and decreasing when above K. It is a constant and must be known before an investigator can calculate the growth of the population from t to t + 1 because K is a parameter of the logistic equation. Unfortunately, for ecologists, K can be not known beforehand, its being simply the mean of the empirically determined series of population sizes (White 1993), making it not particularly useful.

Table 2. Growth of population when 100,000 grams of food are available per day. r_{max} is 0.05, when food available per capita is > 70 g/day. (See text for explanation of table. cap=capita)

Time	N_{UCD}	N_t	Food g/cap	70 − g/cap	Percent difference	ε	r
96	1,429	1,215	82.30	-	-	0.0	0.05
97	1,429	1,277	78.28	-	-	0.0	0.05
98	1,429	1,343	74.47	-	-	0.0	0.05
99	1,429	1,412	70.83	-	-	0.0	0.05
100	1,429	1,484	67.38	2.62	0.1310	0.0066	0.0434
101	1,429	1,550	64.51	5.49	0.2743	0.0137	0.0363
102	1,429	1,607	62.22	7.78	0.3892	0.0195	0.0305
103	1,429	1,657	60.34	9.66	0.4828	0.0241	0.0259
104	1,429	1,701	58.80	11.20	0.5598	0.0280	0.0220
105	1,429	1,738	57.52	12.48	0.6238	0.0312	0.0188

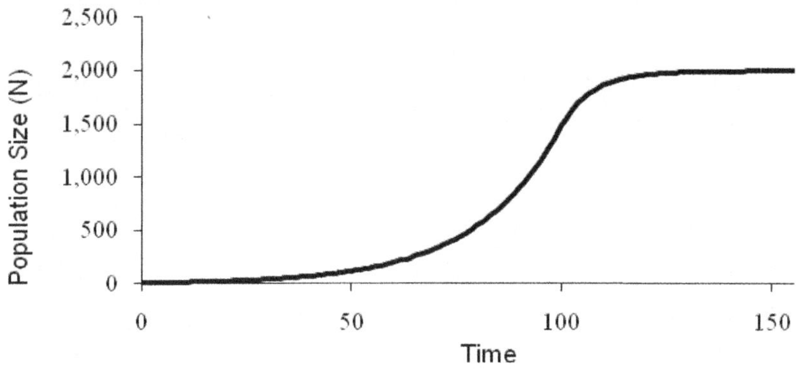

Figure 2. Growth of a food-limited population
when food supply is constant, generated with Eq. 2.

(It seems necessary to point out the caveats. This is a
theoretical study. Its purpose is to show that a population
may grow from a small size to a maximum size beyond
which it does not grow without the intervention of negative
feedback loops [density-dependent factors] and may
continue to for a long time. Again, we start simple and build
to more complexity. What is important here is the model, not
the numbers, which are completely fabricated.)

Variable food supply

In the previous example, the available food supply
was held constant from year to year. Suppose, instead (and
more realistically), that average daily food availability varies
from year to year. In "good" years there is a greater supply,
in "poor" years, a smaller supply. (The numbers are made up
in this simulated example in order for me to make a
quantitative illustration.) For this population, the average
amount of food available per day in each year was
determined from the first 1,250 decimals of the irrational
number, q (Livio 2002), taken five at a time (including four-
digit numbers, such as the 9,179 g/day at t_8 in Table 3, which

corresponds to the decimal numbers 09179, and three-digit numbers, such as 228 at t_{106}, which corresponds to the decimal numbers 00228). Thus, *the mean available daily food supply during each year was independent of population size.*

Suppose, now, that a population with the same characteristics as above—that is, an average individual's probabilities of survival and reproduction are at their maximum values when they consume on average 70 grams of food per capita per day (g/c/d) and that with fewer than 50 (g/c/d), individual survival and reproduction cannot sustain population growth—is living in a variable habitat. Its r_{max} is 0.05. For illustration, I have provided the calculations for years 84 through 93 (Table 3). In year 84 the average food supply is 70,400 g/c/d and $N_{UCD} = 1,006$ (i.e., 70,400/70). At the start of year 84, $N = 435$. Because $N <$ UCD, Eq. 1 applies, $r = r_{max} = 0.05$. $N_{85} = N_{84}e^{0.05} = 458$. Year 85 is particularly poor with an average daily food supply of only 2,812 g/d. Food per capita is 6.14 g/day, which is <70 g/day. This represents a reduction of 3.19 percent of the maximum food consumption per capita—$(70 - 6.14)/20$. Therefore, $\varepsilon = 0.05 \times 0.0319 = 0.1596$, and $r = -0.1096$. Thus, $N_{86} = N_{85}e^{-0.1096} = 410$. I have plotted this population's growth from time 0 ($N_0 = 10$) through time 250 (N_{250}) (Figure 3).

Table 3. Dynamics of a food-limited population with variable food supply. r_{max} is 0.05 when food available per capita is > 70 g/c/day. (See text.)

Time	Food g/day	N_{UCD}	N_t	Food g/capita	70 minus g/capita	Percent difference	ε	R
84	70400	1006	435	161.69	-91.69	-4.58	-0.2292	0.0500
85	2812	40	458	6.14	63.86	3.19	0.1596	-0.1096
86	10427	149	410	25.42	44.58	2.23	0.1115	-0.0615
87	62177	888	386	161.19	-91.19	-4.56	-0.2280	0.0500
88	11177	160	406	27.56	42.44	2.12	0.1061	-0.0561
89	78053	1115	383	203.58	-133.58	-6.68	-0.3339	0.0500
90	15317	219	403	38.00	32.00	1.60	0.0800	-0.0300
91	14101	201	391	36.05	33.95	1.70	0.0849	-0.0349
92	17046	244	378	45.13	24.87	1.24	0.0622	-0.0122
93	66599	951	373	178.47	-108.47	-5.42	-0.2712	0.0500

In this population, the per capita growth rate is a function of the availability of food, being smaller when the amount of food per capita is smaller (Figure 3), and it is not a function of population size (Figure 4). A characteristic of regulated populations is a negative relationship between r and N, per capita growth rate decreasing with increasing population size (e.g., Turchin 2001). Thus, Figure 4 indicates that the population is not regulated, which should not be surprising because regulation was not built into this model of population dynamics.

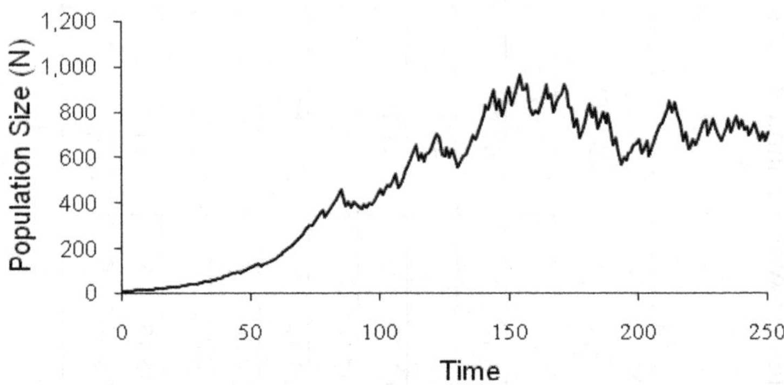

Figure 3. Growth of a population of Species A ($r_{max} = 0.05$) when food supply is variable, predicted with Eq. 2.

When growth is logistic, a plot of ΔN vs. N is characteristically a bell-shaped curve. A plot of ΔN vs. N of Species A is hardly bell-shaped (Figure 4).

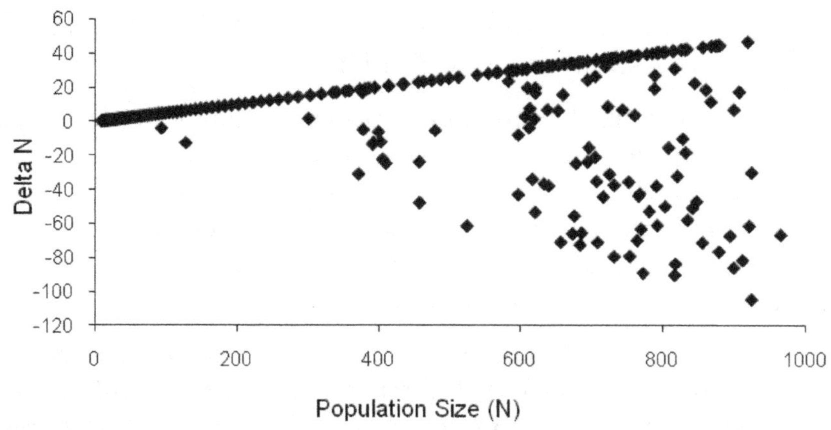

Figure 4. A plot of ΔN vs. N of the population shown in Figure 3.

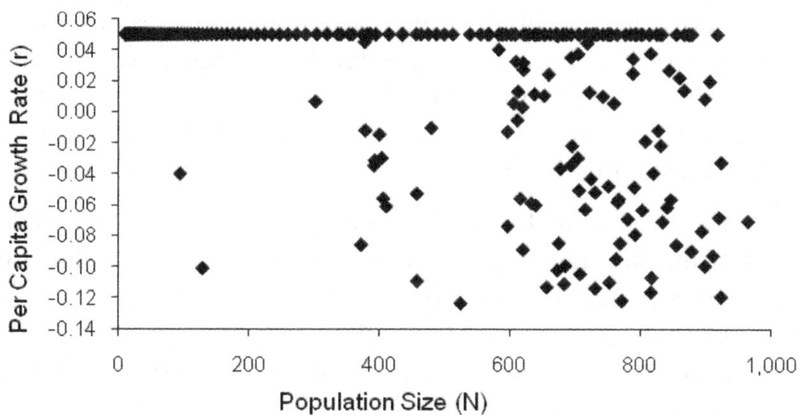

Figure 5. Per capita growth rate (r) plotted against population size (N) of Species A, shown in Figure 3.

When growth is logistic, the growth rate decreases linearly with population size. In Species A, the growth rates below r_{max} seem randomly distributed (Figure 5).

I think the growth of Species A cannot be described as either "logistic" or "density-dependent." This theoretical model constitutes a refutation of the myth promoted by Lack, Royama, Berryman, and many, many others that density-dependent regulation is the *only* plausible model explaining the persistence of populations.

Effect of r_{max}.—Suppose that we have a second species, B, the biology of which is different from Species A with regard to its abilities to acquire, process, and utilize food, such that its maximum growth rate, r_{max}, is greater than that of Species A. Let us compare the growth of Species B with that of Species A by growing Species B in the same environment as we grew Species A in the previous example (Figure 3). With $r_{max} = 0.45$ and with the same variation in food supply as in the previous example, the population undergoes deep fluctuations and reaches a greater N_{max} (Figure 6). As with Species A, the plot of ΔN vs. N is not bell-shaped (Figure 7), and the plot of r vs. N is not negatively linear, or even curvilinear (Figure 8).

There are several features of these examples worth noting. First, the amount of food in any year was independent of previous population size. In this approach to the problem of population dynamics, I model varying environmental conditions—good years and poor years—that affect survival, reproduction, and therefore r. These variations occur independently of population size. Second, population growth in any year is a function of the size of the population at the beginning of the year, the amount of available food, and the individuals' adaptive traits associated with foraging, which determined (i) r_{max}, (ii) the maximum rate of food intake (determining the UCD), (iii) the minimum rate of food intake necessary for sustaining the population (determining N_{max}),

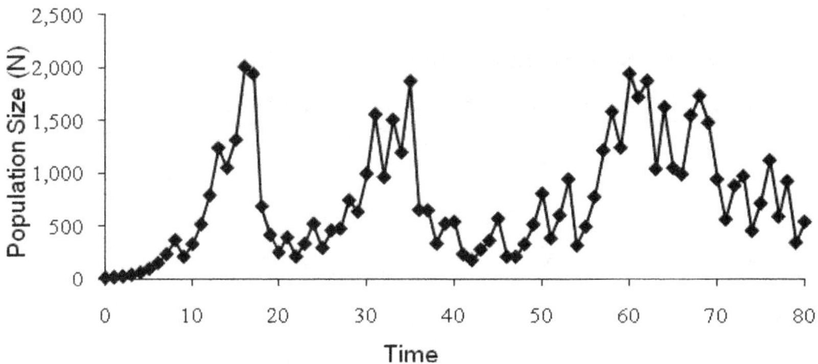

Figure 6. Growth of Species B when food supply is variable, predicted with Eq. 2. The biological properties with regard to the acquisition, processing, or utilization of food of Species B are different from those of Species A shown in Figure 3. The environment (food supply) is exactly the same as in Figure 3.

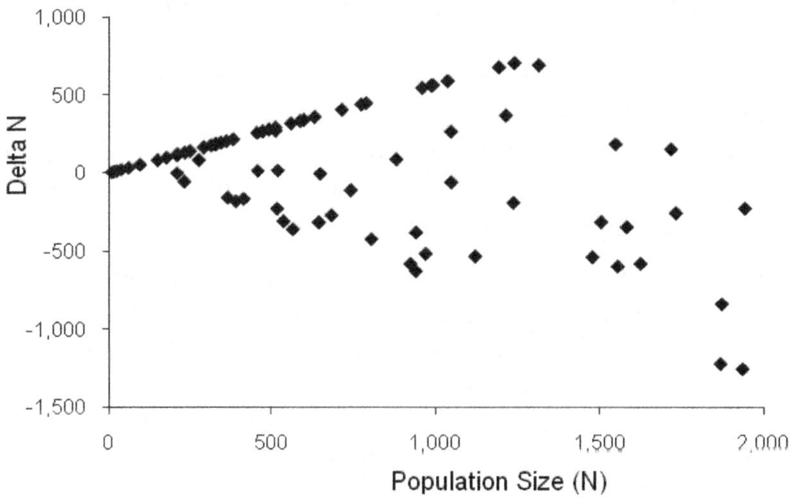

Figure 7. A plot of ΔN vs. N of Species B, shown in Figure 6.

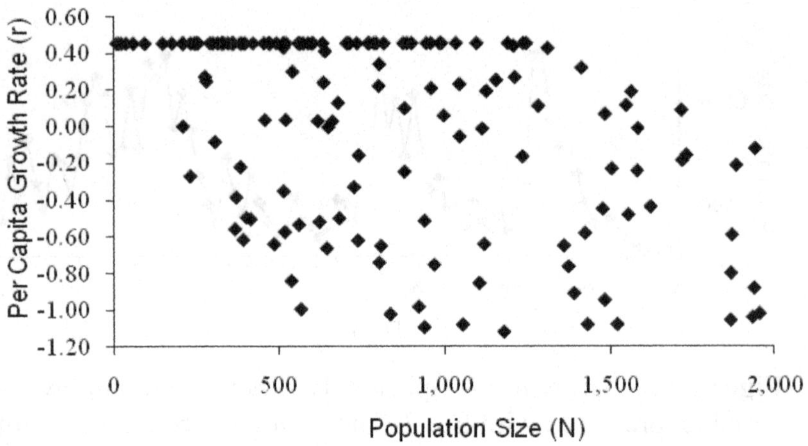

Figure 8. Per capita growth rate (*r*) plotted against population size (*N*) of Species B, shown in Figure 6.

and, therefore, (iv) the range of tolerance. Third, no feedback loop was involved. Nevertheless, the populations were persistent (at least for the 250 years of the one trial calculated for that long). These simulations indicate that populations can fluctuate indefinitely (at least, until extinction), i.e., they can persist, without the aid of density-dependent regulatory mechanisms.

Note that all these graphs were generated with Equation 2, which is much more versatile than the logistic equation for forecasting the growth of populations in time. The logistic can only generate *S*-shaped curves. I dismiss mathematical models invented in other contexts for other purposes because they can hardly be relevant.

A final example

I am including an additional figure, this one for a population with $r_{max} = 0.12$ and food supply held constant in Figure 9.

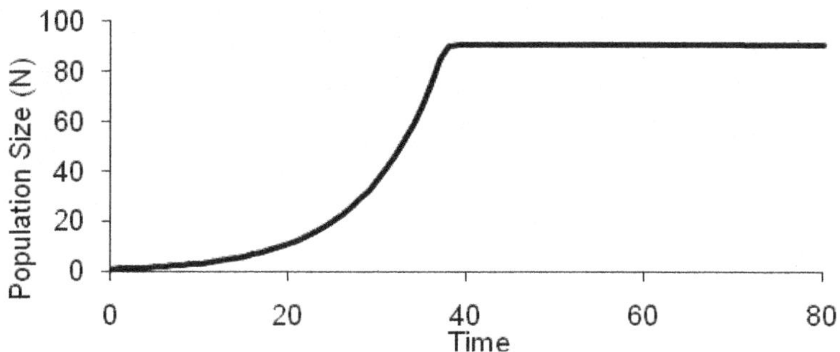

Figure 9. Growth of species C when food supply is constant, predicted with Eq. 2. The biological properties with regard to the acquisition, processing, or utilization of food of Species C are different from those of Species A and B. Its r_{max} is 0.12.

Note that this population is not growing logistically (see pg. 28). Compare this graph with those on *Paramecia* the data for which were obtained from Gause (1934). In his experiments, Gause (1934) attempted to maintain a constant food supply from day to day.

The next step

So far I have considered only those situations in which the food supply was determined independently of population size. The diversity of growth histories "explained" by one simple equation is encouraging to me. The next step is to consider situations in which food supply is affected by population size—producer/herbivore or predator/prey relationships. I modeled predator/prey interactions by subtracting the amount of food consumed by the predator from the prey population prior to the latter's breeding season. The preliminary result was that predator/prey relationships were unstable—the predator eventually eliminated the prey. I next simulated a refuge for a portion of the prey population, which could not be reached by the predator, by subtracting a constant number from the total population size. The predator could feed only on prey not in the refuge. The idea was that the refuge would supply prey to the population outside the refuge. For reasons that I have not resolved, the predator wiped out the prey and became extinct. I probably made a mistake in the model, but I think I was on the right track.

Literature Cited

Berryman, A. A. 1991. Stabilization or regulation: what it all means! Oecologia 86:140- 143.

Dawkins, R. 2006. The God Delusion. Houghton Mifflin, Boston.

Fagerström, T. 1987. On theory, data and mathematics in ecology. Oikos 50:258-261.

Gause, G. F. 1934. *The Struggle for Existence.* Hafner, New York. (reprint 1964).

Lack, D. 1966. Population Studies of Birds. Clarendon Press, Oxford.

Livio, M. 2002. The Golden Ratio. Broadway Books, New York.

Murray, B. G., Jr. 1979. Population Dynamics: Alternative Models. Academic Press, New York.

Murray, B. G., Jr. 1982. On the meaning of density dependence. Oecologia 53:370-373.

Murray, B. G., Jr. 1986. The structure of theory, and the role of competition in community dynamics. Oikos 46:145-158.

Murray, B. G., Jr. 1994. On density dependence. Oikos 69:520-523.

Royama, T. 1977. Population persistence and density dependence. Ecological Monographs 47:1-35.

Sale, P. F., and N. Tolimieri. 2000. Density dependence at some time and place? Oecologia 124:166-171.

Turchin, P. 2001. Does population ecology have general laws. Oikos 94: 17-26.

White, T. C. R. 1993. The Inadequate Environment: Nitrogen and the Abundance of Animals. Springer-Verlag, Berlin.

Chapter 3

Life-history Tables and Life-history Theories

[In his book, *Uncertainty*, David Lindley nicely describes the scientific method: "All sciences aspired to the ideal that physics offered. The trick was to define your science in terms of observations and phenomena that lent themselves to precise description—reducible to numbers, that is—and then to find mathematical laws that tied those numbers into an inescapable system." This is not the way of biologists. Their methods for studying population dynamics (ecology and evolution) are quite different, even though what population biologists study are reducible to numbers (e.g., clutch size, number of clutches laid each year, number of fledglings produced each year, mean number of breeding seasons, number of mates the male or female may have at one time, the number males and females in each age class, number of survivors from year to year, population size, and so on). Instead, in ecology and evolution, we have mathematically oriented investigators who create models of evolutionary change (albeit for simplified populations that have no counterpart in the real world [discussed in the Commentary on Reviews of this paper]) but totally ignore the empirical evidence; that is to say, they do not apply their equations to real situations. Then there are those who love being in the outdoors and collect a plethora of data, but ignore the mathematical theories. There is no attempt (other than my own, dating since 1979) to "find mathematical laws that [tie] those numbers into an inescapable system."

Einstein has said, "The grand aim of all science is to cover the greatest number of empirical facts by logical deduction from the smallest number of hypotheses or axioms." Although anathema to biologists, this is what I attempt to do. I search for universal statements (laws) from which specific predictions can and have been logically deduced. The mind set of biologists is that I cannot be doing what I am doing because what I am doing is "not allowed" in biology.

In 2003 Robert W. Storer created the Katma Award to encourage "the publication and discussion of new ideas, especially those that run counter to established opinion" (see *Condor* 2003: 843). Storer wrote, "Why is katma needed? Science moves forward by the production and acceptance of new ideas, yet it has been increasingly difficult to air new ideas in both pure and applied sciences. Serious work that questions current dogma too often is stifled by those who are angered by seeing their own work questioned." The announcement stated further, "Editors of COS publications will strive not to reject manuscripts simply because they are contrary to established beliefs. They will inform reviewers that they are willing to consider papers with unconventional ideas or innovative approaches as long as these are backed by a well-reasoned argument."

When in August 2003 Joe Jehl told me about the creation of this award, I laughed out loud and told him that the award will never be given because the editors of *The Condor* do not accept papers that are outside the box of conventional wisdom. He assured me that the editors of the Cooper Ornithological Society's journals supported the concept of the Katma Award. Thus, when the official announcement appeared in the November issue, I set about preparing a test case and eventually submitted a manuscript, "The relationship between life-history tables and theories of evolution of life histories." This manuscript was rejected because it presented ideas that were unfamiliar to the reviewers and because it was critical of the conventional wisdom and, by implication, the purveyors of it (including the reviewers).

From what they wrote, the reviewers seemed quite limited in their personal knowledge of the literature on population dynamics, and they seemed limited in their imagination. Nevertheless, on the basis of what I have read during the past decades, I suspected that they represented the views of other ecologists and evolutionary biologists. Their review essentially stated that I was wrong in virtually everything that I thought, and they suggested that if I really wanted to know what life-history tables, life-history theory, and science were all about, I should read books and papers by, for example, Brommer, Caswell, Charlesworth, Cody, McGraw, Ricklefs, Roff, Stearns, and Sutherland. The reviewers did not offer arguments or evidence directly contradicting what I had written. They relied instead on authorities. If the recognized authorities of ecology and evolutionary biology held views contrary to mine, then I must be wrong. There is no need for debate. There is no room for a new theory of population dynamics.

Because of the breadth of the reviewers' criticism, my letter rebutting the reviews evolved into a critique of theoretical ecology and evolutionary biology as they have been practiced during the past forty or so years (regarding the role of population dynamics). In this critique I explain why I did not cite these authors' publications, except perhaps in passing.

In the accompanying Commentary on the reviews, I point out explicitly why I disagree with the "current dogma," as represented by specific, quoted passages from these publications.

Ignored by both theoretical ecologists and field biologists are the Lotka equations in their discrete form. Unfortunately, doing ecology and evolution without using the Lotka discrete equations is like doing physics without Newton's Laws. A real problem is that EEBs (ecologists and evolutionary biologists) have no way of learning the mathematics of population dynamics. It may be hard to believe, but all those textbook descriptions of how to construct life tables are incomplete or opaque, when they are not outright wrong. During the past 30 years I have constructed thousands—I repeat, thousands—of life-history tables. I do not know of another EEB who has made a life-history table outside a classroom or textbook exercise. I do not recall any life-history tables that included life-expectancy at birth or generation time, much less predictions of life-history traits. So, I wrote my paper linking life-history tables to the evolution of life-history traits.

I submitted my paper to the editor of *Condor* on 9 July 2004. This paper was rejected on 23 February 2005, 7½ months after submission. Hearing of my plight and having read my reply to the *Condor*'s reviewers, the editor of the *Auk* invited me to send my paper to him for possible publication as a *Perspective*. This was a perfect place for it to appear, I thought, because my paper gives a very different perspective for the study of life-history evolution. I sent the ms. on 28 December 2005. I have not yet heard one reason why it has not been published.

I include below, the ms. sent to the editor of the *Auk*, revised from that submitted to the *Condor*. The manuscript has been restructured, but the content is unchanged. The dedication (Footnote 1) was in the original. Later footnotes have been added.]

MANUSCRIPT

The Relationship between Life History Tables and Theories of Evolution of Life Histories[1]

Life history tables play little role in discussions of the evolution of life history traits. This is unfortunate because

---

[1] I dedicate this paper to my friend Val Nolan Jr., without whom it could not have been written.

the tables describe the quantitative relationships among population parameters (including age-specific probabilities of surviving and reproducing, age of first breeding, mean expectation of further life, generation time, net reproductive rate, per capita growth rate, and age structure). These relationships must hold in any population, and, consequently, they place constraints on the possible combinations of life history traits. Although the number of possible combinations is unlimited (perhaps a different one for each species), there are even more combinations that are disallowed. Combinations that imply a per capita growth rate $(r) \neq 0$ are disallowed because populations cannot grow to infinite size or become extinct more than once. Thus, such combinations cannot be sustained over evolutionary time.

Because life history traits, such as clutch size, age of first breeding, number of broods reared, mating systems (monogamy, polygyny, polyandry, polygyny-polyandry), sexual dimorphism, and much else, affect or are affected by demographic parameters (e.g. rates of survival and reproduction), and because combinations that result in $r \neq 0$ are disallowed, the life history table is a means of evaluating hypotheses on the evolution of life history traits. Hypotheses proposing disallowed combinations must be rejected.

The premise of this paper is that life history tables provide a foundational structure for testing life history hypotheses. For most ornithologists, ecologists, and evolutionary biologists, however, the construction of a life history table is no more than a classroom or textbook exercise. Life history tables are virtually non-existent in the ornithological literature, and virtually no studies on the evolution of life history traits make reference to life history table information. Knowledge of the construction of life history tables, however, is essential for determining which combinations are allowed and which disallowed. In this paper, I make connections between descriptive life history tables and hypotheses regarding the evolution of the life histories of birds.

THE LIFE HISTORY TABLE

Because textbook treatments showing the construction of life history tables are incomplete, opaque, or just wrong (as will be shown shortly), we must begin with how life history tables should be constructed, in order to determine explicitly the mathematical relationships among demographic parameters. For illustration, we will use a hypothetical population for which we have absolute knowledge (Table 1) This eliminates problems introduced by sampling errors when studying natural populations. With absolute knowledge we will know that our inferences about relationships among demographic parameters will be free of problems introduced by sampling errors. We will know that, for example, as parameter A becomes bigger, parameter B must become smaller, even if we are unable to measure either A or B accurately in a natural population.

Our hypothetical population comprises 15 annual censuses (presume for now that we have counted the females, although counts of males would have been equally suitable). In each column for each time, t_0, t_1, t_2 ..., I have included the number counted alive at time t (age classes one $[n_1]$ through, in this case, six $[n_6]$). The parameter n_0 in the table is the number of (daughter) eggs laid *between* t and $t + 1$ and cannot be counted at t. Nevertheless, the number of eggs laid between t and $t + 1$, i.e., n_0, is entered in the table as $n_{0, t}$. Thus, $n_{0, 14}$ is missing because the census at t_{14} was the last, and $n_{0, 14}$ would have been counted between t_{14} and t_{15}. The population's size (N) at time t is,

$$N_t = \sum_{x=1}^{\infty} n_{x,t} \, , \tag{1}$$

where $n_{x,\,t}$ is the number of individuals of age x at time t. Note that the size of the population in Table 1 and Figure 1, like natural populations, fluctuates in time between an upper and a lower boundary and that the age distribution is not stable (the proportions of age classes do not remain constant from one year to the next).

From the census data, we first calculate age-specific probability of survival (s_x) and age-specific fecundity (m_x). The age-specific probability of surviving from age x at time t to age $x + 1$ at time $t + 1$ (s_x) is calculated with,

$$s_x = \frac{\sum_{t=0}^{\varphi} n_{x+1,\,t+1}}{\sum_{t=0}^{\varphi-1} n_{x,\,t}}, \tag{2}$$

where $n_{x,\,t}$ is the number of individuals of age x at time t, $n_{x+1,\,t+1}$ is the number of individuals of age $x + 1$ at time $t + 1$, and φ is the last census in the sequence.

Bertram G. Murray, Jr.

Table 1. Fifteen censuses of a simulated population. x = age. $n_{x,t}$ is the number of individuals of age x at time t. c_x = proportion of population in age x. b = birth rate. d = death rate. N_t = population size at time t, $\sum_1^\infty n_{x,t}$.

Time	0	1	2	3	4	5	6	7	8	9	10	11	12	13	14		
x	n_x	n_x	n_x	n_x	n_x	n_x	n_x	n_x	n_x	n_x	n_x	n_x	n_x	n_x	n_x	SUM[a]	c_x
0	810	350	270	140	490	80	460	210	610	700	550	380	430	320	--	5800	0.9129
1	18	42	32	22	5	8		29	20	47	11	43	18	29	31	324	0.0486
2	8	5	24	21	8	2	6		28	9	24	1	4	12	16	152	0.0230
3	13	3	1	12	2			5			5	23		1	5	65	0.0096
4	4	1			7								17			29	0.0040
5		2												10		12	0.0016
6			1												1	1	0.0003
Total	853	403	328	195	512	90	466	244	658	756	590	447	469	372			1.0000
N_t	43	53	58	55	22	10	6	34	48	56	40	67	39	52			
b	18.84	6.604	4.655	2.545	22.27	8.00	76.67	6.176	12.71	12.5	13.75	5.672	11.03	6.154			
d	18.60	6.509	4.707	3.145	22.82	8.40	72.00	5.765	12.54	12.79	13.08	6.09	10.69	6.135			
r^b	403/	0.09	-0.05	-0.92	-0.79	-0.51	1.735	0.345	0.154	-0.34	0.516	-0.54	0.288	0.019			
r^c	0.209	0.09	-0.05	-0.92	-0.79	-0.51	1.735	0.345	0.154	-0.34	0.516	-0.54	0.288	0.019			

[a] The sum is over time periods 0 through 13. [b] Calculated with $r = \ln (1 + b - d)$. [c] Calculated with $r = \ln (N_{t+1}/N_t)$.

The mean age-specific fecundity of individuals of age x (m_x) is determined by dividing the number of eggs laid ($n_{0,t}$) from time $t = 0$ to $t = 13$ by the number of birds of the breeding age classes from $t = 0$ to $t = 13$. That is,

$$m_x = \frac{\displaystyle\sum_{t=0}^{\varphi-1} n_{0,t}}{\displaystyle\sum_{t=0}^{\varphi-1}\sum_{x=\alpha}^{\omega} n_{x,t}}, \tag{3}$$

where α is the age of first breeding, and ω is the age of last breeding.

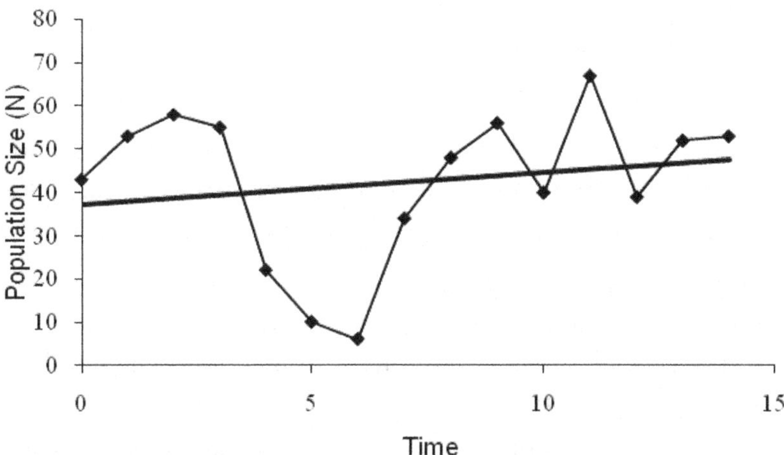

Figure 1. Fluctuations in size of hypothetical population given in Table 1. Correlation: $y = 0.731 + 37.275$, $R^2 = 0.034$.

Many ornithologists will not be able to census their populations in the detail described in the hypothetical example (Table 1). Determining values for s_x and m_x in natural populations is a technical problem best solved by the field investigators. Regardless of how s_x and m_x are determined, the relationship between population parameters calculated from them would be unaffected.

Field ornithologists could calculate age-specific fecundity (m_x) for specific ages x by dividing the number of daughter eggs laid by five-year-old females by the number of five-year-old females. Having determined s_x and m_x, an investigator should proceed as follows.

From survivorship and fecundity data, a life history table may be constructed by calculating, in sequence, l_x, l_xm_x, xl_xm_x, $l_xm_xe^{-rx}$, l_xe^{-rx}, and c_x (Table 2). The probability of surviving from birth (age 0) to a later age x (l_x) is,

$$l_x = \prod_0^{x-1} s_x \, , \, l_0 = 1.0. \tag{4}$$

From having determined values for l_x and m_x for each age x, we may now calculate values for other population parameters. First, calculate l_xm_x for each x and sum; $\sum_0^\infty l_xm_x$, is called the "net reproductive rate" (R_0). It is the mean lifetime reproductive success (in terms of number of female eggs laid/female or, in a male life table, male eggs fertilized/male) of the individuals of a cohort of a population with stable age structure.

Next, calculate xl_xm_x for each x, sum, and divide by R_0; $\sum_0^\infty xl_xm_x \Big/ \sum_0^\infty l_xm_x$, is usually called "generation time" (T), but, more accurately, it is the mean age of breeding of a cohort of males or females of a population with stable age structure. Finally, calculate $\ln(R_0)/T$, the population's per capita growth rate (r). The latter, however, provides only an estimate. The value for r must satisfy the Lotka discrete equation (Birch 1948, Murray and Gårding 1984),

Table 2. Life table for the hypothetical population in Table 1. In this population, $m_x = 9.9485$ (Eq. 3), $R_0 = \sum l_x m_x = 1.0316$, $T = \sum x l_x m_x / \sum l_x m_x = 1.6838$, $r = (\ln R_0)/T = 0.0186$ (corrected such that $1 = \sum l_x m_x e^{-rx}$), $c_x = z l_x e^{-rx}$, and $E = \sum_0^\infty l_x e^{-rx} = 1.1005$. See text for explanation.

Age (x)	s_x	l_x	m_x	$l_x m_x$	$x l_x m_x$	$l_x m_x e^{-rx}$	$l_x e^{-rx}$	c_x
0	0.0581	1.0000	0.0000	0.0000	0.0000	0.0000	1.0000	0.9087
1	0.4938	0.0581	9.9485	0.5780	0.5780	0.5674	0.0570	0.0518
2	0.3750	0.0287	9.9485	0.2855	0.5709	0.2750	0.0276	0.0251
3	0.3846	0.0108	9.9485	0.1070	0.3211	0.1012	0.0102	0.0092
4	0.4138	0.0041	9.9485	0.0412	0.1647	0.0382	0.0038	0.0035
5	0.1667	0.0017	9.9485	0.0170	0.0852	0.0155	0.0016	0.0014
6	0.0000	0.0003	9.9485	0.0028	0.0170	0.0025	0.0003	0.0002
Sum		1.1037		1.0316	1.7370	1.0000	1.1005	1.0000

$$1 = \sum_0^\infty l_x m_x e^{-rx} \,.$$ (5)

This is done by trial and error, by inserting values for r into the equation (easy to do on a spreadsheet) until $\sum_0^\infty l_x m_x e^{-rx}$ equals 1.0000. This degree of precision is greater than field data would allow, but rounding errors should be avoided, especially during calculation and in hypothetical discussions, such as this one, showing the numerical meaning of equations. In field studies, rounding should occur in the reporting of final results.

The mean life expectancy at birth (E) is,

$$E = \sum_0^\infty l_x e^{-rx} \,.$$ (6)

Note that in ecology textbooks $1\big/\sum l_x e^{-rx}$ is said to be the population's birth rate (b), but it is not (see below and Table 2).

The proportion of individuals in age class x in the total population is given by,

$$c_x = z l_x e^{-rx} \,,$$ (7)

where z is $1/E$.

For this population of females, the "net reproductive rate" (R_0) is 1.0316, the "generation time" (T) is 1.6838, the per capita growth rate (r) is 0.0186, life expectancy at birth (E) is 1.1005, and the stable age structure is given in the c_x column (Table 2). These estimates probably closely reflect the mean demographic state of the population over a period of time, even though it never has a stable age structure (Table 1). The life table age structure (c_x column in Table 2) is close to the "empirical" mean age structure (c_x column in Table 1). The life table numbers become closer to the census numbers when longer series of censuses are available (Murray 2000a).

The life history table does not give us the birth rate or death rate. Both are determined directly from the census data. The birth rate between times t and $t + 1$ (b_t) is the number of female (or male) eggs laid ($B_t = n_{0,\,t}$) between times t and $t + 1$ per individual female (or male) alive at time t, that is, $b_t = n_{0,\,t}\Big/\sum_{x=1}^{\infty} n_{x,\,t}$. (Note that the birth rates of males and females differ if the population's sex ratio is not one.) The death rate between times t and $t + 1$ (d_t) is the number of deaths (D_t) of females (or males) between times t and $t + 1$ per individual of females (or males) alive at time t, that is, $d_t = \sum_{x=0}^{\infty}\left(n_{x,\,t} - n_{x+1,\,t+1}\right)\Big/\sum_{x=1}^{\infty} n_{x,\,t}$. (Note that the death rates of males and females differ if the population's sex ratio is not one.) The per capita growth rate between times t and $t + 1$ (r_t) is $\ln(N_{t+1}/N_t)$. The relationship between the birth, death, and growth rates is $r_t = \ln(1 + b_t - d_t)$ (Murray 1997a). These values are shown at the bottom of Table 1.

By working through a life history table we have learned much about the demography of the population. From an input of age-specific survival (s_x) and fecundity (m_x) values, we have calculated values for lifetime reproductive success (*aka* net reproductive rate, R_0), per capita growth rate (r), mean age of breeding (*aka* "generation time," T), mean life expectancy at birth (E), and the proportion of each age class of the population (c_x).

So far, this is a purely descriptive exercise. Life history tables become more useful when we make two further fairly obvious assumptions. First, the Lotka population equations are useful for analysis of populations lacking stable age distributions (Murray 2000a). In other words, a life history table reflects the *average* state of a population over a period of time.

Second, although some populations (a population is a group of individuals of the same species living within a delineated area) do grow rapidly at times or decrease to extinction, most populations fluctuate between upper and lower boundaries with a long-term growth rate of zero during virtually all of their history. For example, suppose the earth's human population grew from 10,000 to 6.3 billion over a period of 100,000 years. The mean per capita growth rate per year would be 0.000134, from $r = \ln(N_t/N_0)/t = \ln(6,300,000,000/10,000)/100,000$—essentially zero. Field data, however, will always provide only an estimate of a population's demographic characteristics, and that only over the period of study. What we can be sure of is that each extant population has had a long period of evolution during which the population's mean per capita growth rate was essentially zero.

Given these assumptions, a study of the life history table shows that any decrease in either s_x or m_x results in a lower r. Any increase in either s_x or m_x results in a greater r. If a population were to have a mean growth rate (r) of zero over evolutionary time, the evolution of any trait that increases (or decreases) s_x or m_x of the population at some age(s) would require a decrease (or increase) in s_x or m_x at some other age or ages. Accordingly, a hypothesis proposing an evolutionary advantage for some trait, one that increases either s_x or m_x over its alternatives, should include an account of how that advantage is balanced by a decrease in some other "component of fitness." *A simple declaration that a trait has been evolved because it increases survival or reproduction is not a complete answer to our understanding of the process of adaptation.* For example, a species that evolved a more cryptic plumage, increasing survival and life expectancy, would grow to infinite size unless some other "component of fitness," such as clutch size, declined. Every positive must be accompanied by a negative; otherwise the population would grow to infinity. Every negative must be accompanied by a positive; otherwise the population would decline to extinction. (Remember that I am describing a method of predicting the relationships among the long-term average values of demographic parameters. I am not describing the day-to-day, month-to-month, or even year-to-year variation in these parameters.)

The solution to this problem is neither easy nor intuitive. In a simulated numerical population (Murray 1997b), the genotype that increased while others decreased did not have the greatest juvenile survivorship, adult survivorship, or fecundity. Nevertheless, testable inferences about the evolution of traits may be made from a rearrangement of Eq. 5 (see below).

Conclusions reached from studying an idealized hypothetical population, like that shown in Table 1, with regard to the relationship between survival and reproduction are unlikely to be changed by the method of determining s_x and m_x. Thus, even without quantitative data, a study of hypothetical life history tables themselves may enlighten life history hypotheses. For example, the postponement of first breeding has sometimes been said to be adaptive because older individuals often are more successful in producing young. Nevertheless, a consideration of the life history table indicates that the postponement of first breeding, say changing from $\alpha = 1$ to $\alpha = 2$, results in $m_1 = 0$ and decreasing r. The cost of this postponement is that the breeding individuals have to lay or fertilize on average more eggs (greater m_x) in every year of reproductive life just to maintain the population's size ($r = 0$) and match the lifetime reproductive success of those individuals starting to breed at an earlier age (Murray 1979, 1985).

NATURAL SELECTION

We usually think about evolution in terms of changes in the frequencies of alternative traits (say, red eyes and white eyes) and their associated genotypes (say, WW, Ww, ww) within a population. These changes are accompanied by changes in each population's life history parameters, as reflected in their having different life history tables. If red eyes has a selective advantage over white eyes because it increases the possessor's probability of surviving or reproducing, then the red-eyed subpopulation will have a different life table from the white-eyed subpopulation. The measure of the rate of change in numbers of a genotype is the Malthusian parameter (Fisher 1930), which is calculated from the genotype's life history table (Murray 1979, 1990, 1995, 1997b).

For clarifying the distinction between the dynamics of a total population of diverse genotypes and the dynamics of its genetically distinct subpopulations, I suggested that we change Eq. 5 to,

$$1 = \sum_x \lambda_x \mu_x e^{-mx} , \tag{8}$$

where m is the Malthusian parameter, λ_x is the probability of an individual with a specific genotype surviving from egg to age x, and μ_x is the mean apparent fecundity of individuals of that genotype and age x (Murray 1979).

Because populations fluctuate between upper and lower boundaries, neither increasing to infinity nor decreasing to extinction, the long-term per capita growth rate (r) of an evolving population is zero. Nevertheless, within a genotypically diverse persisting population (i.e., fluctuating within boundaries with long-term mean $r = 0$), one genotype may be increasing ($m > 0$), while the others are decreasing ($m < 0$), until the genotypes become "fixed," either as a balanced polymorphism or when the alternative genotypes are maintained in the population by mutation, at which time $m = r = 0$ (Murray 1990, 1995, 1997b). The Malthusian parameter is not a fixed number but a variable that changes as a genotype becomes more or less frequent. When a genotype has become fixed, Eq. 5 and Eq. 8 are equivalent.

CLUTCH SIZE EQUATION

From a study of data on the Prairie Warbler (*Dendroica discolor*), Murray and Nolan (1989) proposed an equation for calculating the mean clutch size (C) of a population from other demographic parameters,

$$C = \frac{a+1}{\sum_{\alpha}^{\omega} \lambda_x \sum_{1}^{n} P_i}, \tag{9}$$

where a is the male/female ratio of eggs; λ_x is the probability of survival from egg to age x of individuals with the same genotype, which are the survivors of *successful* clutches (the rearing of at least one young to fledging); P_i is the probability of rearing a brood-i clutch successfully (at least one young leaves the nest); i refers to the first, second, ..., nth brood; $\sum_{\alpha}^{\omega} \lambda_x$ is the mean number of breeding seasons of individuals reared in successful clutches; and $\sum_{1}^{n} P_i$ is the number of broods reared per female during a breeding season. Furthermore, $\sum_{1}^{n} P_i = \sum_{1}^{n} c_i s_i = c_1 s_1 + c_2 s_2 + ... + c_n s_n$, where c_i is the number of brood-i clutches laid per female, and s_i is the probability that a brood-i clutch is successful (Murray 1991a).

The product of $\sum_{\alpha}^{\omega} \lambda_x$ and $\sum_{1}^{n} P_i$, the denominator of Eq. 9, is the mean number of broods reared during a lifetime, that is, lifetime reproductive success, LRS(b). Because evolution maximizes the number of young that are reared from a brood-i clutch (Murray 2006), k_i, lifetime reproductive success in terms of the number of fledglings produced, LRS(k), is $\sum_{\alpha}^{\omega}\lambda_x \sum_1^n P_i k_i$. The genotype with the greatest LRS(b) also has the greatest LRS(k). Thus, I will sometimes refer to lifetime reproductive success as LRS.

THE EVOLUTION OF CLUTCH SIZE

A predictive theory, sometimes called a deductive-nomological (D-N) theory, comprises (a) a set of *universal laws*, which are hypotheses, conjectures, assumptions, or guesses that we *assume* to be universally true for the purpose of analysis and discussion of a problem, (b) a set of *initial conditions*, which are facts that are thought to be true, and (c) the *predictions*, which are singular statements deduced from (a) and (b) (Hempel and Oppenheim 1948, Popper 1968). For example, from his laws and initial conditions (the period and radius of the orbit of the moon), Newton deduced the acceleration of a free-falling body, such as an apple, near the earth's surface (Rogers 1960, Cohen 1980).

Equation 9 accurately predicted the mean clutch size of the three species that have been examined: Prairie Warbler (Murray and Nolan 1989); Florida Scrub-Jay (*Aphelocoma coerulescens*) (Murray et al. 1989); and House Wren (*Troglodytes aedon*) (Kennedy 1991). Furthermore, Wootton et al. (1991) showed that Eq. 9 could be deduced from Eq. 8. This ties the Murray-Nolan clutch-size equation directly to life history tables. Thus, we have no reason to believe that Eq. 9 does not describe the real relationships among a population's demographic parameters.

As just noted, with high quality field data, Eq. 9 makes good predictions of the mean clutch size of populations. It is not necessary to have high quality data, however, in order to deduce testable predictions about the evolution of clutch size. For example, Eq. 9 shows that, in a comparison of two species, if the mean number of breeding seasons ($\sum_{\alpha}^{\omega} \lambda_x$) is greater in one species (because of a greater life expectancy, $\sum_{0}^{\infty} \lambda_x$), then either the mean clutch size (C) or the mean number of broods reared each year ($\sum_{1}^{n} P_i$) must be smaller in that species. Eq. 9 also shows that if parents rear more broods during a breeding season (greater $\sum_{1}^{n} P_i$), the mean clutch size (C) or the mean number of breeding seasons ($\sum_{\alpha}^{\omega} \lambda_x$) must be smaller. Most importantly, Eq. 9 indicates an unexpected relationship between lifetime reproductive success (broods reared; $\sum_{\alpha}^{\omega} \lambda_x \sum_{1}^{n} P_i$), and mean clutch size (C): *As LRS gets larger (or smaller), clutch size must become smaller (or larger).* This important relationship is discussed in detail in the Discussion. Given our assumption that evolving populations are persisting populations (long-term mean $r = 0$), neither increasing to infinity nor becoming extinct, these relationships among parameters must hold.

Life expectancy.—Suppose that life expectancy varies among species. If a single-brooded species is long-lived, then $\sum_{\alpha}^{\omega} \lambda_x$ should be large (depending on α) and clutch size small. If it is short-lived, then $\sum_{\alpha}^{\omega} \lambda_x$ should be small and the clutch size large. This relationship between life expectancy and fecundity is well known (Lack 1954, Saether 1988). According to the view set out in the previous section, natural selection maximizes lifetime reproductive success while minimizing clutch size (Eq. 9). Thus, long-lived species have small clutches because they are long-lived, not because of environmental limitations on their ability to produce eggs or rear young. One should note that this prediction refers to the mean clutch size of a population, not to the size of a particular clutch. Variation about the mean clutch size in a population is an adaptation to differences in environmental quality (Högstedt 1980).

It seems unlikely that life expectancy could evolve as a response to, for example, food availability (Murray 1979, 1985, 1999). Is the small clutch size and long life of albatrosses, vultures, eagles, some parrots, and other organisms the result of a small fecundity imposed by the limitations of food supply (or some other factor) rather than a result of the evolution of relatively large size (Linstedt and Calder 1976, Saether 1989)? I doubt it. Is the large clutch size and short life of phasianids, anatids, and other ground-nesting birds the result of exposure to abundant food supplies rather than to an abundance of predators or other source of mortality? Again, I doubt it.

Length of breeding season.—Suppose that breeding seasons vary in length, say longer in the tropics and shorter at higher latitudes. If so, $\sum_1^n P_i$ should be large when breeding seasons are long because each female has a prolonged period within which to replace clutches that fail, resulting in a high c_1 and, thus, a high P_1, and because, if a species is multibrooded, the long breeding season allows the rearing of second and third broods (high P_2 and P_3). Thus, populations with long breeding seasons, as in the tropics, should tend to have large $\sum_1^n P_i$ and small clutch sizes. When the breeding season is short, as at high latitudes, fewer, if any, clutches may be laid following clutch failure (small c_1) and second and third broods are rare to nonexistent (small P_2 and P_3). Thus, when the breeding season is short, $\sum_1^n P_i$ should be small and the clutch size large. Because the breeding seasons tend to be longer at low latitudes and shorter at high latitudes, we should expect to see an increase in clutch size with increasing latitude, a well-known phenomenon in birds (Lack 1947, 1948, Klomp 1970, Cody 1971, Böhning-Gaese et al. 2000). The prediction, however, is that clutch size is causally related to the length of the breeding season, not with latitude. Thus, a tropical species with a short breeding season should have a small $\sum_1^n P_i$ and large clutch size. The Yellow-throated Euphonia (*Euphonia hirundinacea*) in Costa Rica, for example, has a large clutch size (four, compared with the frequent two in tropical populations [Skutch 1985]) but is single-brooded during its short breeding season (Sargent 1993), three months compared with the average of six months for many other species in Costa Rica (Ricklefs 1969). We should expect that a temperate zone species, such as the Mourning Dove (*Zenaida macroura*), with a long breeding season and many broods (mean of three successful broods per year; McClure 1942, Nice 1957), should have large $\sum_1^n P_i$ and a small clutch size (Murray 1985, 1991b). Other seemingly anomalous geographic variations in clutch-size are consistent with this hypothesis (Murray, ms.).

Predation.—Suppose that the intensity of predation and, thus, life expectancy ($\sum_0^\infty \lambda_x$) and, therefore, $\sum_\alpha^\omega \lambda_x$ vary among populations. If predation is high, then $\sum_\alpha^\omega \lambda_x$ should be small, and the mean clutch size large. If predation is low, then $\sum_\alpha^\omega \lambda_x$ should be large, and the mean clutch size small. This relationship was shown in the most thorough comparative study undertaken: clutch size in the Parulinae is larger and predation rate greater in ground-nesting species than in above-ground-nesting species, among genera within the subfamily, among species within a genus, among subspecies within a species, and among individuals within a subspecies (Martin 1988). The hypothesis that predation on eggs or chicks selects for small clutch size (Skutch 1949, Perrins 1977, Slagsvold 1982, Martin 1992) is inconsistent with both the facts and the prediction of the theory proposed here.

In addition to these examples based on mathematical arguments, the benefits of a small clutch size outweigh the benefits of a large clutch size (Murray 1999). Because the breeding season is a finite time period, the advantages of a smaller clutch size are (i) an increase in the probability that any young will leave a clutch, (ii) a greater number of broods reared in multibrooded species, (iii) the greater likelihood that a female will lay a replacement clutch after having laid a small clutch than a large clutch (Slagsvold 1984, Smith et al. 1987, Tinbergen 1987, Lindén 1988), (iv) the time interval between consecutive clutches is shorter with smaller than with larger clutches (McGillivray 1983, Slagsvold 1984, Smith et al. 1987), (v) better protection of young that have left the nest (Safriel 1975), and (vi) the production of heavier young, which may increase the survival of young through their first year (Perrins 1965, Lack 1966, von Haartman 1971, Murphy 1978, Perrins 1980, Garnett 1981). The only benefit of a larger clutch is the greater number of fledglings reared per *successful* clutch.

LIFE HISTORY EQUATION

We can write Eq. 9 for both males and females in the following manner because the mean clutch size of a population must be the same for both sexes,

$$\frac{a+1}{\sum_{\alpha}^{\omega}\lambda_x\sum_{1}^{n}P_i}(for\ males)=\frac{a+1}{\sum_{\alpha}^{\omega}\lambda_x\sum_{1}^{n}P_i}(for\ females).$$

(10)

A change in one parameter on one side of the equation implies a change on the other side or a second change on the first side. A single change is disallowed. The life history table (Table 2) was for the females of a population. The males, too, have a life history table, which could be quite different from the females' because of differences in age-specific survival and fecundity and age of first breeding. Furthermore, although different, the male and female life history tables must have the same per capita growth rate (r)—otherwise one sex would be increasing more rapidly than the other (Murray 1990).

A useful relationship is obtained by rearranging Eq. 9,

$$F_r = C\sum\nolimits_1^n P_i = \frac{a+1}{\sum\limits_\alpha^\omega \lambda_x} , \qquad (11)$$

where F_r is the replacement fecundity, that is, the annual mean number of eggs laid per female or fertilized per male of a persisting population ($r = 0$) during a breeding season (Murray 1979). If F_r is greater than C, then $\sum\nolimits_1^n P_i$ must be greater than one, and the population must be multibrooded. Multibroodedness may occur with pairs rearing more than one brood in a monogamous relationship (they may switch mates between broods—i.e., serial monogamy), if the breeding season is sufficiently long, or with each member of a pair tending separate broods simultaneously, as in the Red-legged Partridge (*Alectoris rufa*) (Green 1984) and Mountain Quail (*Oreortyx pictus*) (Beck et al. 2005), or by one mate obtaining multiple mates in polygamous relationships. Furthermore, if we signify the replacement fecundity (F_r) of populations with monogamous relationships with F_m, then polygamous relationships should occur when $F_r > F_m$. Although these equations are descriptive of demographic relationships, they are helpful in understanding the evolution of mating systems.

THE EVOLUTION OF MATING SYSTEMS

Let us consider how Eqs. 10 and 11 can be useful in understanding the evolution of mating systems (referring to social organization—monogamy, polygyny, polyandry, and polygyny-polyandry—rather than to such behaviors as extra-pair copulations). It should be remembered that the predictions refer to the combination of life history parameters to be expected in a particular relationship. The theory does not necessarily explain how the particular combination evolved. For example, high mortality may come about because of factors such as a high incidence of predation, or disease, or aggression, a particularly severe climate, or something else. A predictive theory is not the end of research but the beginning, suggesting new avenues to be explored.

Monogamy.—When both sexes are monogamous during each breeding episode (from the laying of the first egg through independence of young), even though they may switch mates between broods, we know that $\sum_1^n P_i (\male) = \sum_1^n P_i (\female)$. Thus, $\sum_\alpha^\omega \lambda_x (\male) = \sum_\alpha^\omega \lambda_x (\female)$. The sexes, however, may differ in life expectancy ($\sum_0^\infty \lambda_x$) and in age of first breeding (α). If males survive better (or worse) than females, their age of first breeding should be later (or earlier) than that of females. In many passerine species, there seems to be an excess of males, as indicated by the often rapid replacement of males that are removed from their territories (e.g., Hensley and Cope 1951, Stewart and Aldrich 1951). We should expect that males would have a later age of first breeding. For example, in the monogamous Florida Scrub-Jay, life expectancy of males is greater than in females (because of greater mortality of females during the first year of life), and males begin breeding at a later age (Woolfenden and Fitzpatrick 1984). In some species, however, females live longer than males, and, therefore, $\alpha(\female)$ should be later than $\alpha(\male)$. In the single-brooded, monogamous Laysan Albatross (*Phoebastria immutabilis*), for example, females survive better than males and begin breeding at a later age

(Fisher 1975a). Thus, monogamy may occur regardless of which sex is more numerous. A condition for monogamy is $F_r = F_m = C\sum_1^n P_i$ for both males and females (Table 3).

Table 3. The relationship between replacement fecundity ($F_r = C\sum_1^n P_i$), calculated with Eq. 11, and the annual fecundity that may be expected from monogamous relationships (determined empirically) in relation to mating system. See text for discussion.

Male/female	Male	Female
Monogamous/monogamous	$F_m = C\sum_1^n P_i$	$F_m = C\sum_1^n P_i$
Polygynous/monogamous	$F_m < C\sum_1^n P_i$	$F_m = C\sum_1^n P_i$
Monogamous/polyandrous	$F_m = C\sum_1^n P_i$	$F_m < C\sum_1^n P_i$
Polygynous/polyandrous	$F_m < C\sum_1^n P_i$	$F_m < C\sum_1^n P_i$

These examples emphasize that Eq. 10 predicts patterns of demographic parameters rather than provide final explanations for the patterns. Why do males survive better than females in the Florida Scrub-Jay but less well in the Laysan Albatross?

Polygyny.—When males have two or more mates with overlapping breeding episodes and females have only one mate, we know that on average the number of broods reared/male is greater than the number of broods reared/female, i.e., $\sum_1^n P_i (\male) > \sum_1^n P_i (\female)$. Thus, $\sum_\alpha^\omega \lambda_x (\male)$

should be smaller than $\sum_{\alpha}^{\omega}\lambda_x\,(\female)$. A smaller $\sum_{\alpha}^{\omega}\lambda_x\,(\male)$ could occur with either a shorter life expectancy ($\sum_{0}^{\infty}\lambda_x\,(\male)$) or a later age of first breeding (α). Polygyny may occur regardless of the adult sex ratio because, if males outnumber females, $\sum_{\alpha}^{\omega}\lambda_x\,(\male)$ could always be made smaller than $\sum_{\alpha}^{\omega}\lambda_x\,(\female)$ by males' postponing age of first breeding (α) through aggressive or other dominance behavior. If instead males survive less well than females, then M/F < 1 and $\sum_{\alpha}^{\omega}\lambda_x\,(\male)$ is smaller than $\sum_{\alpha}^{\omega}\lambda_x\,(\female)$. Thus, theoretically, polygyny may occur regardless of which sex is more numerous (Murray 1984). The theory tells us that the conditions for polygyny are $F_r > F_m$ for males and $F_r = F_m$ for females (Table 3), and that $\sum_{\alpha}^{\omega}\lambda_x\,(\male)$ should be smaller than $\sum_{\alpha}^{\omega}\lambda_x\,(\female)$. Again, there is much to be learned. We are still left to determine why polygyny takes so many forms (as described by Emlen and Oring 1977, Wittenberger 1981, and Oring 1982).

Polyandry.— Polyandry is a rare mating relationship among birds (Lack 1968, Oring 1986), the evolution of which still seems to remain a puzzle (Clutton-Brock 1991, Ligon 1999, Bennett and Owens 2002, Andersson 2005). A demographic theory of its evolution (Murray 1984) seems to have been overlooked. Because polyandry is rare, we may assume that it evolved from some ancestral population, which was either monogamous or polygynous and in which the males were usually aggressive in sequestering access to females. The evolutionary problem is to explain the change from a population of aggressive, often territorial, males to one of aggressive, often territorial females. When females have two or more mates with overlapping breeding episodes and males have only one mate, we know that on average the number of broods/female is greater than the number of broods/male, i.e., $\sum_1^n P_i\,(\female) > \sum_1^n P_i\,(\male)$. Thus, $\sum_\alpha^\omega \lambda_x\,(\female)$ should be smaller than $\sum_\alpha^\omega \lambda_x\,(\male)$. A smaller $\sum_\alpha^\omega \lambda_x\,(\female)$ could occur with either a shorter life expectancy ($\sum_0^\infty \lambda_x\,(\female)$) or a later age of first breeding (α). The latter seems unlikely, especially if adult females outnumber or are equal to the number of males (M/F $\leq$ 1). The evolution of aggression in females that postponed first breeding would result in a population including unmated but fertile females while males were limited to sharing the polyandrous mates. We would have to find a selective advantage for males to evolve away from being aggressive and having one or more monogamous mates to sharing mates with other males while unmated females were still around. Even if one could imagine a selective advantage for such a change in behavior, females would be limited in the number of clutches they could lay for mates, whereas males are able to fertilize the eggs of more females than females are able to lay clutches for males.

Alternatively, a reduction in $\sum_\alpha^\omega \lambda_x\,(\female)$ could easily occur by a reduction in life expectancy. No mutation that reduces life expectancy would likely be selected, but suppose that some factor (as yet unknown) reduces life expectancy of females, resulting in M/F > 1. With a sufficient shortage of females, males could no longer be polygynous, or even monogamous. If males waited to mate with an unmated female, however, $\alpha(\male)$ would increase and $\sum_\alpha^\omega \lambda_x\,(\male)$ would decrease, requiring a greater $\sum_1^n P_i\,(\male)$ when there are no females available for mating.

According to this hypothesis, females are polyandrous because males usually have no opportunity to be either monogamous or polygynous (except incidentally).

The conditions for polyandry are M/F > 1, $F_r = F_m$ for males, and $F_r > F_m$ for females (Table 3). This theoretical discussion is better understood by example.

Consider the Spotted Sandpiper (*Actitis macularia*), which is polyandrous, has a four-egg clutch, and a male can rear no more than one brood in a breeding season (Oring et al. 1997). If females were monogamous, they could rear only one brood from one four-egg clutch per year ($F_r = F_m = 4$). If $F_r(♀) > 4$, then $\sum_1^n P_i\,(♀) > 1$, and because males can rear no more than one brood per year, females must be polyandrous and $\sum_\alpha^\omega \lambda_x\,(♀) < \sum_\alpha^\omega \lambda_x\,(♂)$. Because both sexes begin breeding at age 1 (Oring et al. 1997), mortality must be greater in females, from which one must infer that the adult M/F ratio should be > 1.

These demographic considerations are what led me to predict (oral paper, 1980 AOU meeting) that the adult M/F ratio in the Spotted Sandpiper should be > 1, when it was believed to be ≈ 1 (Oring 1982). Oring et al. (1983) confirmed this prediction, reporting that M/F = 1.39. Thus, in 1980, the prediction of M/F > 1 in the Spotted Sandpiper was a genuine prediction from theory of something unknown by anyone at the time.

High quality data on survival rates would be invaluable for rigorously testing the predictions of theory, but they are essentially lacking. Some limited but suggestive comparative data, however, are available for the polyandrous Spotted Sandpiper (Oring et al. 1983, Reed and Oring 1993, Oring et al. 1997) and the monogamous Common Sandpiper (*Actitis hypoleucos*) (Holland et al. 1982). Survival from one year to the next of adult Common Sandpipers is much greater (90%) than of adult Spotted Sandpipers (about 60%). The probability of rearing a clutch to hatching is greater in the Common Sandpiper (89%) than in the Spotted Sandpiper (about 44%). Both these figures imply a greater $\sum_{\alpha}^{\omega}\lambda_x$ in the Common Sandpiper. Because both species have a clutch size of four eggs, Eq. 10 tells us that $\sum_{1}^{n}P_i$ should be greater in the Spotted Sandpiper, that is it should produce more broods each year than the Common Sandpiper does. In neither species do we have sex-specific survival data. The prediction from theory, then, is that $F_r = F_m$ in both sexes of the Common Sandpiper and the male Spotted Sandpiper, but $F_r > F_m$ in the female Spotted Sandpiper.

Three conditions seem necessary for the evolution of polyandry: (i) greater survival of males than females, resulting in adult M/F ratio > 1, (ii) a male alone is capable of rearing a brood, and (iii) high mortality in females, resulting in $F_r(♀) > F_m(♀)$ (Table 3). I do not consider it coincidental that polyandrous species are characterized by having M/F > 1 (Murray 1984). In order to test the theory more rigorously, however, we need some high quality data on survival of males and females.

Again, this theory only predicts the relationships among demographic parameters that characterize different mating relationships. In order to understand the evolution of polyandry, we should want to know why mortality is greater in females than in males. Why is mortality so high in the Spotted Sandpiper compared with the congeneric Common Sandpiper? We should also want to know why polyandry is usually sequential in some species (e.g., Spotted Sandpiper) and simultaneous in other species with males either tending separate clutches (e.g., Northern Jacana [*Jacana spinosa*]; Jenni and Collier 1972) or tending a single clutch (e.g., Galapagos Hawk [*Buteo galapagoensis*]; de Vries 1975, Faaborg et al. 1980).

Polygyny-polyandry.—An initial condition for monogamy, polygyny, and polyandry is that $F_r = F_m$ for at least one of the sexes. Suppose that mortality is so great that

$F_r > F_m$ for both sexes. If so, each sex must be polygamous. Then, the relationship is polygyny-polyandry (Table 3), which takes a variety of forms. Among them, there is double-clutching with mate-switching, as in the Mountain Plover (*Charadrius montanus*) (Graul 1973) and Temminck's Stint (*Calidris temminckii*) (Hildén 1975), with each member of the pair caring for a brood. In some species, some females may take a third mate, implying (a prediction) that on average males are more numerous than females.

A very different relationship is found in the Greater Rhea (*Rhea americana*), Boucard Tinamou (*Crypturellus boucardi*), and Brushland Tinamou (*Nothoprocta cinerascens*)—a group of females mates with a male and deposits the eggs in his nest before moving on to another male (Lancaster 1964a, b, Bruning 1974). Females neither incubate eggs nor care for the young. It has been suggested that this reproductive pattern provides a male with a full clutch of eggs in a short time (Lancaster 1964b, Orians 1969, Wiley 1974).

More recent work has shown that the reproductive behavior of male rheas is much more complicated, with some males incubating only, others copulating only, and others both copulating and incubating (Martella et al. 1998). Nevertheless, the question remains, Why should the males' clutches be so large in the first place? My explanation is that mortality of both males and females is so great that $\sum_{\alpha}^{\omega}\lambda_x$ is small and $F_r > F_m$ for both sexes. I agree that this reproductive pattern provides a male with a large "clutch" of eggs in a short time, but the pattern itself evolves because of a high predation rate on eggs and, perhaps, chicks.

A unique mating system is that of the Common Ostrich (*Ostrich camelo*) (Bertram 1992). A male establishes a territory and pairs with a major female. The clutch size is large, about 22 eggs, half of which are laid by the major female, the other half by one to more than five minor females. Most of the major female's eggs are fertilized by the territorial male, but most of the minor females' eggs are fertilized by other males. Only the major female participates in incubation and care of the young. Minor females copulate with several males, and lay their eggs in several nests. Why should such an unusual system evolve? Females outnumber males, perhaps favoring polygyny, but ostriches are long-lived, perhaps living 30 years in nature (Bertram 1992). The Laysan Albatross life history is similar in that males outnumber females, both sexes are long-lived, and a pair rears no more than one brood per year, but $C = 1$. Why is reproduction of albatrosses and ostriches so different? According to Eq. 9 and 10, if $\sum_1^n P_i \leq 1$ and C large, as in the ostrich, then $\sum_\alpha^\omega \lambda_x$ must be small, which can come about by high mortality in the pre-breeding ages. Indeed, mortality of ostrich nests is very high, perhaps one in 100 eggs producing a yearling (Bertram 1992), whereas about 35% of albatross eggs produce a seven-year old (Fisher 1975b). Thus, $\sum_\alpha^\omega \lambda_x$ is much larger in albatrosses than in ostriches. In contrast to rheas, successful reproduction of ostriches may require two parents to incubate because untended eggs could overheat during mid-day departures from the nest by an adult for feeding or nest defense (Bertram 1992).

These few examples show how Eqs. 9, 10, and 11 can be used to interpret the evolution of life histories. The number of combinations is unlimited. As I have indicated before (Murray 1991a, 1999, 2000b), the required data for analyses are virtually unreported in the literature. This is not because ornithologists do not have them but because they do not arrange or report their field data in such a way as to allow calculation or comparison.[2]

---

[2] I believe this to be so because field investigators lack an understanding of the life table and therefore the relationships of the life table parameters.

DISCUSSION

Equation 9 reveals an important relationship—an inverse relationship between mean clutch size (C) and LRS ($\sum_{\alpha}^{\omega}\lambda_x \sum_1^n P_i$). Any mutation that increases life expectancy by increasing age-specific survival at one or more ages should itself be selected for and increase in frequency. Its Malthusian parameter (m) would be greater than zero. As the new mutation increases in frequency, the population's r would be greater than zero, and the population would increase. Of course, a population cannot increase indefinitely. At first environmental factors would reduce survival or fecundity, preventing further growth. Eventually, mutations for a smaller clutch size should be favored for the reasons given above. Thus, a population of longer-lived, less fecund organisms evolves from a short-lived, more fecund ancestor.

We should not expect any mutation that increases mortality and reduces lifetime reproductive success to evolve because the mutant genotype would have a lower m than the wild genotype. If environmental change, however, such as increased predation, results in greater mortality and a lowering of LRS, a larger clutch size could be selected for. Thus, a population of shorter-lived, more fecund individuals could evolve from a longer-lived, less fecund population. Whether it does or does not, however, depends on whether a larger clutch size is within the natural variation of the genotype. Often, however, an environment that results in greater mortality and lower LRS should cause extinction of the population.

These considerations led me to suggest the following law: *The fittest life history has the greatest lifetime reproductive success.* It may seem counterintuitive that the greatest lifetime reproductive success is associated with a lowered clutch size, but that is the clear implication of Eqs. 9 and 10. This rule is consistent with what I have called the third law of evolution, *Selection favors those females that lay as few eggs or bear as few young as are consistent with replacement because they have the highest probability of surviving to breed again, their young have the highest probability of surviving to breed, or both* (Murray 1979, 1985, 1991b, 2001) . Scientific laws do not tell us what is true. They tell us what may be true, and until they are shown to be false we cannot know that they are untrue. Laws are part of a deductive-nomological theory and, therefore, are testable (Hempel and Oppenheim 1948, Hempel 1965, Popper 1968, 1979).

Laws do not predict or explain everything that we want to know. Once we understand, for example, that the Spotted Sandpiper is polyandrous because females are too short-lived to sustain the population by producing a single brood in each year of reproductive life, we should want to know why females are so short-lived, especially compared with its congener, the Common Sandpiper. Why do females consistently survive less well than males over evolutionary time? On this point, despite its importance to our understanding of the evolution of polyandry, there has not even been speculation.

Biologists are correct. The biology of plants and animals is extraordinarily complex. The environment provides a myriad of initial conditions resulting in great biological diversity. What we must do is figure out which set of initial conditions is responsible for what empirical fact.

With regard to clutch size, there are many species that have clutches of the same size, and the laws do not explain why that should be. The explanation lies in the diversity of initial conditions. Clutch size is a function of (i) life expectancy ($E = \sum_0^\infty \lambda_x$), which is itself a consequence of age-specific survival (s_x), (ii) the age of first breeding (α), and (iii) the probabilities of rearing one, two, or more broods (P_1, P_2, etc.), which are themselves a function of the number of clutches laid per female per brood (c_1, c_2, etc.) and the probabilities of rearing any young from a brood-i clutch (s_1, s_2, etc.). Two species could differ in any two (or more) of these and have the same clutch size, and these differences could be brought about by many different environmental factors. The combinations seem unlimited. In any particular case, we should want to know whether species A has a greater clutch size than species B because it has, as initial conditions, a shorter breeding season, lower daily nest survival, later age of first breeding, shorter life expectancy, or something else, or several differences. This can only be determined empirically.

Furthermore, two species could differ in any of these and have different clutch sizes. Then, we should want to know why the species differ in survival or age of first breeding or whatever. For example, Why does the Yellow-throated Euphonia have such a short breeding season in the tropics? Ultimately, we would want to know why one species differs from another, and this cannot be determined with deductive-nomological (D-N) theory. D-N theory, however, allows us to separate allowed explanations from disallowed explanations.

The D-N model is not a means for producing the final solutions to our problems. Rather it allows us to evaluate our hypotheses, to suggest lines of research, and to produce genuinely testable non-ad hoc hypotheses. The prediction that the Spotted Sandpiper is polyandrous because males outnumber females was testable. The prediction that the large clutch size of the tropical Yellow-throated Euphonia is a result of a short breeding season was testable.

Philosophy.—According to Einstein, "The grand aim of all science is to cover the greatest number of empirical facts by logical deduction from the smallest number of hypotheses or axioms" (Barnett 1950). Insofar as biology has no recognized predictive theories, that is, theories comprising a small number of universal laws and the quantitative predictions that can be logically deduced from them, one might infer that biological theories are not scientific (Murray 2001). Prominent biologists, including Mayr (1982, 1988, 1996), Snow (1985), Bartholomew (1986), Slobodkin (1988), and Lawton (1999), have contended that universal laws are not to be expected because of the complexity of biological systems and the historical nature of evolution. It seems a truism, however, that scientists who believe that universal laws cannot be found will not find them (Murray 1975, Jehl and Murray 1986). Universal laws can only be found by an active and dedicated search for them because they are creations of the human mind, not something that can be found empirically (Bronowski 1965). Furthermore, as Popper (1979:193) has pointed out, "Only if we require that explanations shall make use of universal statements or laws of nature (supplemented by initial conditions) can we make progress towards realizing the idea of independent, or non-*ad hoc*, explanations."

Among biologists, ornithologists have the best opportunity to discover universal laws because birds are better known in many ways than any other large taxon. Thus, despite the views of many biologists that biological diversity is an impediment to the discovery of predictive theory, I consider that it offers opportunity for the discovery and testing of universal laws. The well described patterns of variations in clutch size, mating systems, and sexual size dimorphism (to mention a few) are "regularities," and regularities should have some kind of explanation beyond simple ad hoc hypotheses. Should biologists not consider that regularities in nature may follow from some small set of universal rules?

The purpose of this paper is to explore this possibility in a way that has not been done before. By linking life history table information (e.g., survival and reproduction) to the life history traits of birds (e.g., clutch size and mating systems), I show that empirical facts are predictable from statements presumed to be universally true.

Biological diversity does not necessarily preclude a limited number of biological processes, such as natural selection, being universally true. Instead of asking themselves, "What ad hoc hypothesis would explain my observations," investigators might better ask, "If natural selection is universally true, what initial conditions would lead to the evolution of the biological facts that I know?" In this way, we biologists may join Einstein and other scientists in the search for the smallest number of hypotheses or axioms that cover the greatest number of empirical facts.

ACKNOWLEDGMENTS

I wish to thank J. R. Jehl, Jr., J. Burger, and M. Gochfeld for their thoughtful recommendations, which improved my presentation of a difficult subject.

Literature Cited

ANDERSSON, M. 2005. Evolution of classical polyandry: three steps to female emancipation. Ethology 111:1-23.

BARNETT, L. 1950. The meaning of Einstein's new theory. Life 28:22.

BARTHOLOMEW, G. A. 1986. The role of natural history in contemporary biology. BioScience 36:324-329.

BECK, J. L., K. P. REESE, P. ZAGER, AND P. E. HEEKIN. 2005. Simultaneous multiple clutches and female breeding success in Mountain Quail. Condor 107:889-897.

BENNETT, P. M., AND I. P. F. OWENS. 2002. Evolutionary Ecology of Birds: Life Histories, Mating Systems, and Extinction. Oxford University Press, Oxford.

BERTRAM, B. C. R. 1992. The Ostrich Communal Nesting System. Princeton University Press, Princeton, NJ.

BIRCH, L. C. 1948. The intrinsic rate of natural increase of an insect population. Journal of Animal Ecology 17:15-26.

BÖHNING-GAESE, K., B. HALBE, N. LEMOINE, AND R. OBERRATH. 2000. Factors influencing the clutch size, number of broods and annual fecundity of North American and European land birds. Evolutionary Ecology Research 2:823-839.

BRONOWSKI, J. 1965. Science and Human Values. revised ed. Harper & Row, New York.

BRUNING, D. F. 1974. Social structure and reproductive behavior in the Greater Rhea. Living Bird 13:251-294.

CLUTTON-BROCK, T. H. 1991. The Evolution of Parental Care. Princeton University Press, Princeton, NJ.

CODY, M. L. 1971. Ecological aspects of reproduction. Pages 461-512 *in* Avian Biology, vol I (D. S. Farner and J. R. King, eds.). Academic Press, New York.

COHEN, I. B. 1980. The Newtonian Revolution. Cambridge University Press, Cambridge.

DE VRIES, T. 1975. The breeding biology of the Galapagos Hawk, *Buteo galapagoensis*. Gerfaut 65:29-54.

EMLEN, S. T., AND L. W. ORING. 1977. Ecology, sexual selection, and the evolution of mating systems. Science 197:215-223.

FAABORG, J., T. DE VRIES, C. B. PATTERSON, AND C. R. GRIFFIN. 1980. Preliminary observations on the occurrence and evolution of polyandry in the Galapagos Hawk (*Buteo galapagoensis*). Auk 97:581-590.

FISHER, H. I. 1975a. The relationship between deferred breeding and mortality in the Laysan Albatross. Auk 92:433-441.

FISHER, H. I. 1975b. Mortality and survival in the Laysan Albatross, *Diomedea immutabilis*. Pacific Science 29:279-300.

FISHER, R. A. 1930. The Genetical Theory of Natural Selection. Clarendon Press, Oxford.

GARNETT, M. C. 1981. Body size, its heritability and influence on juvenile survival among Great Tits, *Parus major*. Ibis 123:31-41.

GRAUL, W. D. 1973. Adaptive aspects of the Mountain Plover social system. Living Bird 12:69-94.

GREEN, R. E. 1984. Double nesting of the Red-legged Partridge *Alectoris rufa.* Ibis 126:332-346.

HEMPEL, C. G. 1965. Aspects of Scientific Explanation and other Essays in the Philosophy of Science. The Free Press, New York.

HEMPEL, C. G., AND P. OPPENHEIM. 1948. Studies in the logic of explanation. Philosophy of Science 15:135-175.

HENSLEY, M. M., AND J. B. COPE. 1951. Further data on removal and repopulation of the breeding birds of a spruce-fir forest community. Auk 68:483-493.

HILDÉN, O. 1975. Breeding system of Temminck's Stint *Calidris temminckii.* Ornis Fennica 52:117-146.

HÖGSTEDT, G. 1980. Evolution of clutch size in birds: adaptive variation in relation to territory quality. Science 210:1148-1150.

HOLLAND, P. K., J. E. ROBSON, AND D. W. YALDEN. 1982. The breeding biology of the Common Sandpiper *Actitis hypoleucos* in the Peak District. Bird Study 29:99-110.

JEHL, J. R., JR., AND B. G. MURRAY, JR. 1986. The evolution of normal and reverse sexual size dimorphism in shorebirds and other birds. Pages 1-86 *in Current Ornithology,* vol 3 (R. F. Johnston, ed.) Plenum, New York.

JENNI, D. A., AND G. COLLIER. 1972. Polyandry in the American Jacana (*Jacana spinosa*). Auk 743:743-765.

KENNEDY, E. D. 1991. Predicting clutch size of the House Wren with the Murray-Nolan equation. Auk 108:728-731.

KLOMP, H. 1970. The determination of clutch-size in birds. A review. Ardea 58:1-124.

LACK, D. 1947. The significance of clutch-size. Parts I and II. Ibis 89:302-352.

LACK, D. 1948. The significance of clutch-size. Part III. Ibis 90:25-45.

LACK, D. 1954. The Natural Regulation of Animal Numbers. Oxford University Press, London.

LACK, D. 1966. Population Studies of Birds. Clarendon Press, Oxford.

LACK, D. 1968. Ecological Adaptations for Breeding in Birds. Methuen, London.

LANCASTER, D. A. 1964a. Biology of the Brushland Tinamou, *Nothoprocta cinerascens*. Bull. Am. Mus.Nat. Hist. 127:273-314.

LANCASTER, D. A. 1964b. Life history of the Boucard Tinamou in British Honduras. Part II: Breeding biology. Condor 66:253-276.

LAWTON, J. H. 1999. Are there general laws in ecology? Oikos 84:177-192.

LIGON, J. D. 1999. The Evolution of Avian Breeding Systems. Oxford University Press, Oxford.

LINDÉN, M. 1988. Reproductive trade-off between first and second clutches in the Great Tit *Parus major*: an experimental study. Oikos 51:285-290.

LINDSTEDT, S. L., and W. A. CALDER. 1976. Body size and longevity in birds. Condor 78:91-94.

MARTELLA, M. B., J. L. NAVARRO, AND L. BELLIS. 1998. New evidence on the mating system of *Rhea americana*. Ostrich 69:259-260.

MARTIN, T. E. 1988. Nest placement: implications for selected life-history traits, with special reference to clutch size. American Naturalist 132:900-910.

MARTIN, T. E. 1992. Interaction of nest predation and food limitation in reproductive strategies. Pages 163-197 *in* Current Ornithology, vol 9 (D. M. Power, ed. Plenum, New York.

MAYR, E. 1982. The Growth of Biological Thought: Diversity, Evolution, and Inheritance. Harvard University Press, Cambridge, MA.

MAYR, E. 1988. Toward a New Philosophy of Biology. Harvard University Press, Cambridge, MA.

MAYR, E. 1996. The autonomy of biology: the position of biology among the sciences. Quarterly Review of Biology 71:97-106.

McCLURE, H. E. 1942. Mourning Dove production in southwestern Iowa. Auk 59:64-75.

McGILLIVRAY, W. B. 1983. Intraseasonal reproductive costs for the House Sparrow (*Passer domesticus*). Auk 100:25-32.

MURPHY, E. C. 1978. Seasonal variation in reproductive output of House Sparrows: the determination of clutch size. Ecology 59:1189-1199.

MURRAY, B. G., JR. 1975. Distinguishing the woods from the trees, or seeking unity in diversity. BioScience 25:149.

MURRAY, B. G., JR. 1979. Population Dynamics: Alternative Models. Academic Press, New York.

MURRAY, B. G., JR. 1984. A demographic theory on the evolution of mating systems as exemplified by birds. Pages 71-140 *in Evolutionary Biology*, vol 18 (M. K. Hecht, B. Wallace, and G. T. Prance, eds.). Plenum, New York.

MURRAY, B. G., JR. 1985. Evolution of clutch size in tropical species of birds. Pages 505-519 *in* Neotropical Ornithology (P. A. Buckley, M. S. Foster, E. S. Morton, R. S. Ridgely, and F. G. Buckley, eds.). American Ornithologists' Union, Washington, D.C.

MURRAY, B. G., JR. 1990. Population dynamics, genetic change, and the measurement of fitness. Oikos 59:189-199.

MURRAY, B. G., JR. 1991a. Measuring annual reproductive success, with comments on the evolution of reproductive behavior. Auk 108:942-952.

MURRAY, B. G., JR. 1991b. Sir Isaac Newton and the evolution of clutch size: a defense of the hypothetico-deductive method in ecology and evolutionary biology. Pages 143-180 *in* Beyond Belief: Randomness, Prediction, and Explanation in Science (J. Casti and A. Karlqvist, eds.). CRC Press, Boca Raton, FL.

MURRAY, B. G., JR. 1995. A method for projecting genotypic change in populations with complex genetic and demographic structure. Oikos 73:415-418.

MURRAY, B. G., JR. 1997a. On calculating birth and death rates. Oikos 78:384-387.

MURRAY, B. G., JR. 1997b. Population dynamics of evolutionary change: demographic parameters as indicators of fitness. Theoretical Population Biology 51:180-184.

MURRAY, B. G., JR. 1999. Predicting the occurrence of synchronous and asynchronous hatching in birds. Pages 624-637 *in* Proceedings of the 22nd International Ornithological Congress, vol 22 (N. J. Adams and R. H. Slotow, eds.). BirdLife South Africa, Durban, South Africa.

MURRAY, B. G., JR. 2000a. Dynamics of an age-structured population drawn from a random numbers table. Austral Ecology 25:297-304.

MURRAY, B. G., JR. 2000b. Measuring annual reproductive success in birds. Condor 102:470-473.

MURRAY, B. G., JR. 2001. Are ecological and evolutionary theories scientific? Biological Reviews 76:255-289.

MURRAY, B. G., JR. 2006. A new equation for calculating reproductive success of clutches as a function of the day on which incubation starts: some implications. Auk 123:708-721.

MURRAY, B. G., JR., AND L. GÅRDING. 1984. On the meaning of parameter x of Lotka's discrete equations. Oikos 42:323-326.

MURRAY, B. G., JR., AND V. NOLAN, JR. 1989. The evolution of clutch size. I. An equation for predicting clutch size. Evolution 43:1699-1705.

MURRAY, B. G., JR., J. W. FITZPATRICK, AND G. E. WOOLFENDEN. 1989. The evolution of clutch size. II. A test of the Murray-Nolan equation. Evolution 43:1706-1711.

NICE, M. M. 1957. Nesting success in altricial birds. Auk 74:305-321.

ORIANS, G. H. 1969. On the evolution of mating systems in birds and mammals. American Naturalist 103:589-603.

ORING, L. W. 1982. Avian mating systems. Pages 1-92 *in Avian Biology*, vol 6 (D. S. Farner, J. R. King, and K. C. Parkes, eds.). Academic Press, New York.

ORING, L. W. 1986. Avian polyandry. Pages 309-351 *in* Current Ornithology, vol 3 (R. F. Johnston, ed. Plenum, New York.

ORING, L. W., E. M. GRAY, AND J. M. REED. 1997. Spotted Sandpiper (*Actitis macularia*) *in* The Birds of North America, vol 289 (A. Poole and F. Gill, eds.). Academy of Natural Sciences, American Ornithologists Union, Philadephia, PA, Washington, D. C.

ORING, L. W., D. B. LANK, AND S. J. MAXSON. 1983. Population studies of the polyandrous Spotted Sandpiper. Auk 100:272-285.

PERRINS, C. M. 1965. Population fluctuations and clutch-size in the Great Tit, *Parus major* L. Journal of Animal Ecology 34:601-647.

PERRINS, C. M. 1977. The role of predation in the evolution of clutch size. Pages 181-191 *in* Evolutionary Ecology (B. Stonehouse and C. Perrins, eds.). University Park Press, Baltimore, MD.

PERRINS, C. M. 1980. The survival of young Great Tits *Parus major*. Proceedings of the International Ornithological Congress 17:159-174.

POPPER, K. R. 1968. The Logic of Scientific Discovery. revised ed.. Harper & Row, New York.

POPPER, K. R. 1979. Objective Knowledge: An Evolutionary Approach (revised ed.). Oxford University Press, London.

REED, J. M., AND L. W. ORING. 1993. Philopatry, site fidelity, dispersal, and survival of Spotted Sandpipers. Auk 110:541-551.

RICKLEFS, R. E. 1969. The nesting cycle of songbirds in tropical and temperate regions. Living Bird 8:165-175.

ROGERS, E. M. 1960. Physics for the Inquiring Mind: The Methods, Nature, and Philosophy of Physical Science. Princeton University Press, Princeton, NJ.

SAETHER, B.-E. 1988. Pattern of covariation between life-history traits of European birds. Nature 331616-617.

SAETHER, B.-E. 1989. Survival rates in relation to body weight in European birds. Ornis Scandinavica 20:13-21.

SAFRIEL, U. N. 1975. On the significance of clutch size in nidifugous birds. Ecology 56:703-708.

SARGENT, S. 1993. Nesting biology of the Yellow-throated Euphonia: large clutch size in a neotropical frugivore. Wilson Bulletin 105:285-300.

SKUTCH, A. F. 1949. Do tropical birds rear as many young as they can nourish? Ibis 91:430-455.

SKUTCH, A. F. 1985. Clutch size, nesting success, and predation on nests of neotropical birds, reviewed. Pages 575-594 *in* Neotropical Ornithology (P. A. Buckley, M. S. Foster, E. S. Morton, R. S. Ridgely, and F. G. Buckley, eds.). American Ornithologists' Union, Washington, D.C.

SLAGSVOLD, T. 1982. Clutch size variation in passerine birds: the nest predation hypothesis. Oecologia 54:159-169.

SLAGSVOLD, T. 1984. Clutch size variation of birds in relation to nest predation: on the cost of reproduction. Journal of Animal Ecology 53:945-953.

SLOBODKIN, L. B. 1988. Intellectual problems of applied ecology. BioScience 38:337-342.

SMITH, H. G., H. KÄLLANDER, AND J.-Å. NILSSON. 1987. Effect of experimentally altered brood size on frequency and timing of second clutches in the Great Tit. Auk 104:700-706.

SNOW, D. W. 1985. Review of "Sexual selection, lek and area behavior and sexual size dimorphism in birds" by R. Payne. Auk 102:231-233.

STEWART, R. E., AND J. W. ALDRICH. 1951. Removal and repopulation of breeding birds in a spruce-fir community. Auk 68:471-482.

TINBERGEN, J. M. 1987. Costs of reproduction in the Great Tit: intraseasonal costs associated with brood size. Ardea 75:111-122.

VON HAARTMAN, L. 1971. Population dynamics. Pages 391-459 *in* Avian Biology, vol 1 (D. S. Farner and J. R. King, eds.). Academic Press, New York.

WILEY, R. H. 1974. Evolution of social organization and life-history patterns among grouse (Aves: Tetraonidae). Quarterly Review of Biology 49:201-227.

WITTENBERGER, J. F. 1981. Animal Social Behavior. Duxbury Press, Boston.

WOOLFENDEN, G. E., AND J. W. FITZPATRICK. 1984. The Florida Scrub Jay:

Demography of a Cooperative-breeding Bird. Princeton University Press, Princeton, NJ.

WOOTTON, J. T., B. E. YOUNG, AND D. W. WINKLER. 1991. Ecological versus evolutionary hypotheses: demographic stasis and the Murray-Nolan clutch size equation. Evolution 45:1947-1950.

Chapter 4

Demography and Population Dynamics of the Ivory-billed Woodpecker

[When I first heard of the (now alleged) rediscovery of the Ivory-billed Woodpecker, I was as excited as anyone. Perhaps, I still had a chance of seeing one. As a population ecologist, it was not long before I asked myself, how many are there? The bird was either a member of an extant population, or the last of a few birds of a viable population that had remained hidden in the swamps of the south-central states until recently. The likelihood that it was a lone very old bird roaming around the southern swamps for a half century seems most unlikely (i.e., "annual probability of survival" raised to the power "number of years" becomes a very small number, that is, highly improbable). So, I modeled several simulated populations of "Ivory-billed Woodpeckers," on the basis of informed guesses of annual survival rates, mean clutch size, and age of first breeding, in order to see what a small *persisting* population would, or should, look like.

With an eye for detail, ecologists and evolutionary biologists (EEBs) will automatically assume that this paper is about the Ivory-billed Woodpecker, which it is, and therefore it will be of little interest to those uninterested in Ivory-billed Woodpecker biology. In fact, for a theoretical person, such as myself, it is also about the population dynamics of small populations—*any* small population. What is the smallest viable population for any species of interest? Put another way, What is the minimum mean size of a persisting population? In this paper, I propose a method for estimating a minimum mean size of a viable population.

I submitted this paper to the editor of the *Auk* on 24 April 2006. It was rejected on 2 July 2006. Unfortunately, ecologists have little understanding of life history tables, much less what can be inferred from them. I sent my commentary on the reviews to the editor on 6 July 2006.]

MANUSCRIPT

Demography and Population Dynamics of the Ivory-billed Woodpecker: The Dynamics of Small Populations

ABSTRACT.—The recent report of a sighting of the Ivory-billed Woodpecker raises several questions. If it was a genuine sighting, a question becomes whether this is the last of the species or representative of a small persisting population? Even if the last individual, a small population must have survived until recently. What then would a small persisting population of Ivory-billed Woodpeckers look like? From reasonable assumptions about the life history of the Ivory-billed Woodpecker (mean clutch size of three eggs, breeding beginning at age two, and maximum age between 15 and 20 years) and from use of demographic equations, I suggest that the minimum mean size of a persisting population should be about 30 post-juvenile birds.

***

The reported Ivory-billed Woodpecker (*Campephilus principalis*) sighting in Arkansas (Fitzpatrick et al. 2005) raises several questions regarding the possibility that a small population has survived in the south-central United States essentially undetected during the past half century. If the sighting was of an Ivory-billed Woodpecker, and this is by no means certain (Fitzpatrick et al. 2006, Jackson 2006, Sibley et al. 2006), is this bird the last survivor of the species or a member of a persisting small population? This same question could have been asked about any of the reported sightings of the past fifty years. Whether it is the last bird can only be determined by noting whether an Ivory-billed Woodpecker is observed again ten, twenty, or fifty years from now. Even if this is the last bird, it would be the last of a small population that persisted until recently, and if a small population persisted until a decade or so ago, then the species' final extinction was caused by recent events rather than by logging in the first half of the 20th Century.

What, then, would be the demographic characteristics of a persisting small population of Ivory-billed Woodpeckers? Unfortunately, we have no direct information, other than clutch and brood sizes. Bendire (1895, cited in Allen and Kellogg 1937, Bent 1939) reported clutches of three to five eggs, but Bent (1939) himself thought that the clutch did not normally exceed three. The number of young that fledged was usually only one, but there are two reports of pairs with four young (Bent 1939) and one with five nestlings (Allen and Kellogg 1937). Thus, observations are few and quite variable. Jackson (2004:29) summarized the data, "Reported clutch sizes range from one to six eggs and brood sizes from one to four young." Also Jackson (2002:17-18) noted, "Three of 6 nests Tanner (1941) found in the Singer Tract were successful in producing 1 young each. ... Between 1931 and 1939, at least 19 young in 9 broods were observed out of the nest, averaging 2.11 young/brood (Tanner 1941, Tanner [1942])." With 100 percent fledging success of eggs in these nests, the mean clutch size must have been greater than two. For the sake of discussion, then, let us assume that the mean clutch size of the Ivory-billed Woodpecker is (or was) three eggs.

Even though we must acknowledge that we can never know the demography of the Ivory-billed Woodpecker, we can consider the plausible possibilities. The purpose of this study is not to determine its demography precisely but to consider what a small population of Ivory-billed Woodpeckers would look like.

POPULATION EQUATIONS

Let us begin with an equation that gives the mean size ($\overline{N}$) of any persisting population (neither increasing to infinite size nor decreasing to extinction—i.e., the long-term per capita growth rate [r] is zero),

$$\overline{N} = \overline{n}_0 \sum_0^\infty l_x - \overline{n}_0 \, , \tag{1}$$

where $\overline{n}_0$ is the mean initial size of cohorts and $\sum_0^\infty l_x$ is the mean life expectancy at birth (Murray 2003). All we need to know are $\overline{n}_0$ and $\sum_0^\infty l_x$ in order to estimate the mean size of a persisting population. Note, however, that a population's size N at any time t is,

$$N_t = \sum_1^\infty n_{x,t} \, , \tag{2}$$

where $n_{x,t}$ is the number of individuals of age x (>0) at time t.

A fundamental population equation is Lotka's equation in its more useful discrete form (Birch 1948, Murray and Gårding 1984),

$$1 = \sum l_x m_x e^{-rx} \, , \tag{3}$$

where l_x is the probability of an individual's surviving from birth to age x, m_x is the number of daughters per female (or sons per male) of age x, e is the base of the natural logarithms, and r is the per capita rate of change.

There is no theoretical method for determining the mean initial size of cohorts—$\overline{n}_0$ can only be determined empirically, by counting the number of eggs laid each year over a period of time. In the hypothetical discussion of a minimum $\overline{N}$ of a small persisting population, then, we can only compare the consequences of assuming different values of $\bar{n}_0$—say, 9, 18, 27, 36, and 45 eggs.

$\sum_0^\infty l_x$ is determined from a population's life history table. Lacking hard evidence, a life history table of the Ivory-billed Woodpecker must be based on several assumptions. First, the long-term per capita growth rate, r, is zero (and, therefore, $\sum_0^\infty l_x m_x = 1.0$); that is, the population is persisting (r and N from year to year, however, may vary). Second, if females lay only a single clutch per year, the mean annual fecundity (m_x) is one half the mean clutch size (C) (because the life history table is based on one sex). If birds relay after a clutch failure, as the Ivory-billed Woodpecker occasionally did (Jackson 2002), m_x would be a bit larger, but a persisting population with greater m_x would have a smaller survival rate during the first year or later or both, and thus would have no effect on the *conclusions* of this exercise. Third, an expected maximum age (ϖ) must be assumed. The maximum age so far reported of the other large North American woodpecker, the Pileated Woodpecker (*Dryocopus pileatus*), is 12 years, 11 months (K. Klimkiewicz, pers. comm.), and the lifespan of "large woodpeckers rarely exceed 15 years" (Fitzpatrick et al. 2005). Jackson (2004), however, suggested that the Ivory-billed Woodpecker was longer-lived, perhaps having a "natural potential longevity of 20-plus years." Fourth, the probability of any individual surviving to its potential maximum age should be small. (For example, the probability of a Prairie Warbler [*Dendroica discolor*] reaching its maximum known age of 12 years, from being laid as an egg, is 0.0007 [Nolan 1978, Murray and Nolan 1989]. That for a Florida Scrub-Jay [*Aphelocoma coerelescens*] reaching its maximum known age of 15 years, from being laid as an egg, is 0.0077 [Woolfenden and Fitzpatrick 1984, Murray et al. 1989].)

Requiring that a persisting population's life history be limited to a given age and a given mean clutch size limits the number of plausible life histories that can be imagined.

HYPOTHETICAL LIFE HISTORY TABLES

For illustrative purposes, I have constructed a life history table (Table 1) for a persisting population (which I will call Population A), with a maximum individual age of nineteen years and a mean fecundity (m_x) of 1.65 eggs, assuming a mean clutch size of three eggs and ten percent of females relaying after failure. Because non-breeding yearlings and families foraging and traveling together have been reported by several observers (Jackson 2004), I assumed an age of first breeding (α of 2. I adjusted adult age-specific survival ($s_{x>0}$) to be a constant 0.73, and the survival rate from egg to age one (s_0) to be 0.225, such that $r = 0.00005$, essentially zero. With these numbers the probability of reaching a maximum age of 19 years (l_{19}) is 0.0008. (In working with life history tables, it is best to "estimate" numerical parameters to three, four, or five decimal places so that small but important numerical differences do not disappear and because there are several ways to cross-check the calculations precisely. We can always round off our final numbers.)

One can imagine combinations of numbers for other populations (Table 2). For example, keeping s_0 and α unchanged, one can imagine a population with a shorter (B) or a longer (C) maximum life expectancy. All that is necessary, for example, is to decrease (to 0.60 for Population B) or increase (to 0.79 for Population C) adult age-specific survival.

I have made life history tables (not shown here) for two other populations, keeping the probability of surviving to maximum age ($l_\varpi = 0.0008$) and first-year survival ($s_0 =$

0.225) the same as in Population A, and adjusting adult survival ($s_{x>0}$), fecundity (m_x), and age of first breeding (α), such that $r = 0$. The shorter-lived population (D) and longer-lived population (E) are quite different from each other and from Population A (Table 2). Population D has lower adult survival (lower $s_{x>0}$), greater m_x, earlier age of first breeding (α), and obviously a shorter life expectancy at birth ($\sum_{0}^{\infty} l_x$). Population E has greater adult survival (greater $s_{x>0}$), smaller m_x, later age of first breeding (α), and a greater life expectancy at birth (Table 2). (I could have also changed first-year survival [s_0], but I tried to keep the number of varying variables as small as possible.)

In our illustrative life history (Population A in Tables 1 and 2), $\sum_{0}^{\infty} l_x$ is 1.8312. From Eq. 1 we can calculate $\overline{N}$ (ages > 0). With initial size of cohort, $\overline{n}_0 = 9$ (mean number of eggs laid per year), the mean population size ($\overline{N}$) is 7.5 birds (including non-breeding yearlings); with $\overline{n}_0 = 18$, $\overline{N} = 15$; with $\overline{n}_0 = 27$, $\overline{N} = 22.4$; with $\overline{n}_0 = 36$, $\overline{N} = 29.9$; and with $\overline{n}_0 = 45$, $\overline{N} = 37.4$.

A persisting population with $\alpha = 1$ and $m_x = 1.782$ (population D in Table 2) has $\sum_{0}^{\infty} l_x$ of 1.5613. With $\overline{n}_0 = 9$, $\overline{N} = 5.5$ birds; with $\overline{n}_0 = 18$, $\overline{N} = 10.1$; with $\overline{n}_0 = 27$, $\overline{N} =$ unchanged, one can imagine a population with a shorter (B) or a longer (C) maximum life expectancy. All that is necessary, for example, is to decrease (to 0.60 for Population B) or increase (to 0.79 for Population C) adult age-specific survival.

Table 1. Life history table of hypothetical population A (s_x, age-specific probability of surviving from age x to age $x + 1$; l_x, the probability of surviving from birth to age x; m_x, mean age-specific fecundity; $\sum l_x m_x$, per capita lifetime reproductive success [*aka* net reproductive rate, R_0]; $\sum l_x m_x e^{-rx}$ must equal 1.0 [Lotka 1925], determined by trial and error; $\sum l_x e^{-rx}$ is life expectancy at birth, E_0; c_x is the proportion of the population comprised of individuals of age x).

Age (x)	s_x	l_x	m_x	$l_x m_x$	$x l_x m_x$	$l_x m_x e^{-rx}$	$l_x e^{-rx}$	c_x
0	0.225	1.0000	0.00	0.0000	0.0000	0.0000	1.0000	0.5461
1	0.730	0.2250	0.00	0.0000	0.0000	0.0000	0.2250	0.1229
2	0.730	0.1643	1.65	0.2710	0.5420	0.2710	0.1642	0.0897
3	0.730	0.1199	1.65	0.1978	0.5935	0.1978	0.1199	0.0655
4	0.730	0.0875	1.65	0.1444	0.5777	0.1444	0.0875	0.0478
5	0.730	0.0639	1.65	0.1054	0.5271	0.1054	0.0639	0.0349
6	0.730	0.0466	1.65	0.0770	0.4618	0.0769	0.0466	0.0255
7	0.730	0.0341	1.65	0.0562	0.3933	0.0562	0.0340	0.0186
8	0.730	0.0249	1.65	0.0410	0.3281	0.0410	0.0248	0.0136
9	0.730	0.0181	1.65	0.0299	0.2695	0.0299	0.0181	0.0099
10	0.730	0.0132	1.65	0.0219	0.2186	0.0218	0.0132	0.0072
11	0.730	0.0097	1.65	0.0160	0.1755	0.0159	0.0097	0.0053
12	0.730	0.0071	1.65	0.0116	0.1398	0.0116	0.0071	0.0039
13	0.730	0.0052	1.65	0.0085	0.1105	0.0085	0.0051	0.0028

14	0.730	0.0038	1.65	0.0062	0.0869	0.0062	0.0038	0.0021
15	0.730	0.0027	1.65	0.0045	0.0680	0.0045	0.0027	0.0015
16	0.730	0.0020	1.65	0.0033	0.0529	0.0033	0.0020	0.0011
17	0.730	0.0015	1.65	0.0024	0.0410	0.0024	0.0015	0.0008
18	0.730	0.0011	1.65	0.0018	0.0317	0.0018	0.0011	0.0006
19	0.000	0.0008	1.65	0.0013	0.0244	0.0013	0.0008	0.0004
Sum		1.8312		1.0003	4.6424	1.0000	1.8311	1.0000

$R_0 = 1.0003$, $T = 4.6411$, $r = 0.00005$, $E_0 = 1.8312$.

I have made life history tables (not shown here) for two other populations, keeping the probability of surviving to maximum age ($l_\varpi = 0.0008$) and first-year survival ($s_0 = 0.225$) the same as in Population A, and adjusting adult survival ($s_{x>0}$), fecundity (m_x), and age of first breeding (α), such that $r = 0$. The shorter-lived population (D) and longer-lived population (E) are quite different from each other and from Population A (Table 2). Population D has lower adult survival (lower $s_{x>0}$), greater m_x, earlier age of first breeding (α), and obviously a shorter life expectancy at birth ($\sum_0^\infty l_x$). Population E has greater adult survival (greater $s_{x>0}$), smaller m_x, later age of first breeding (α), and a greater life expectancy at birth (Table 2). (I could have also changed first-year survival [s_0], but I tried to keep the number of varying variables as small as possible.)

In our illustrative life history (Population A in Tables 1 and 2), $\sum_0^\infty l_x$ is 1.8312. From Eq. 1 we can calculate $\overline{N}$ (ages > 0). With initial size of cohort, $\overline{n}_0 = 9$ (mean number of eggs laid per year), the mean population size ($\overline{N}$) is 7.5 birds (including non-breeding yearlings); with $\overline{n}_0 = 18$, $\overline{N} = 15$; with $\overline{n}_0 = 27$, $\overline{N} = 22.4$; with $\overline{n}_0 = 36$, $\overline{N} = 29.9$; and with $\overline{n}_0 = 45$, $\overline{N} = 37.4$.

A persisting population with $\alpha = 1$ and $m_x = 1.782$ (population D in Table 2) has $\sum_0^\infty l_x$ of 1.5613. With $\overline{n}_0 = 9$, $\overline{N} = 5.5$ birds; with $\overline{n}_0 = 18$, $\overline{N} = 10.1$; with $\overline{n}_0 = 27$, $\overline{N} = 15.2$; with $\overline{n}_0 = 36$, $\overline{N} = 20.2$; and with $\overline{n}_0 = 45$, $\overline{N} = 25.3$. A persisting population with $\alpha = 3$ and $m_x = 1.502$ (population E in Table 2) has $\sum_0^\infty l_x$ of 2.0685. With $\overline{n}_0 = 9$, $\overline{N} = 9.6$ birds (including non-breeding yearlings and second-year birds); with $\overline{n}_0 = 18$, $\overline{N} = 19.2$; with $\overline{n}_0 = 27$,

$\overline{N}$ = 28.8; with $\overline{n}_0$ = 36, $\overline{N}$ = 38.5; and with $\overline{n}_0$ = 45, $\overline{N}$ = 48.1. (I have not shown the life history tables for Populations B, C, D, and E, but I have supplied enough information [Table 2] for others to construct their own and check my results for themselves.)

For these populations, we may also determine (i) R_0, the lifetime reproductive success of a cohort (i.e., number of daughters per female born to a cohort), otherwise known as "net reproductive rate," with $\sum l_x m_x$, (ii) T, the generation time with $\sum x l_x m_x / \sum l_x m_x$, and (iii), E_0, the mean life expectancy at birth with $\sum_0^\infty l_x$. The demography of the five hypothetical populations is compared in Table 2. Populations A, B, and C are virtually indistinguishable, despite their having different maximum ages. This is because s_0, $s_{x>0}$, and m_x were kept the same in each population while their maximum ages were arbitrarily chosen (the probability of an individual reaching maximum age [l_ω] is different in these populations). Populations A, D, and E are quite different because s_0, $s_{x>0}$, m_x, and α had to be adjusted to accommodate the assumptions that the maximum age in each population had to be reached with equal probability and m_x had to be kept close to 1.65.

POPULATION DYNAMICS OF SMALL POPULATIONS

The c_x column of a life history table gives the stable age structure of a population—the proportion in the population of individuals of age x. The age structure of each of the populations determined from the five life history tables is given in Table 2. We can convert these proportions to numbers in each age class, assuming a given $\overline{n}_0$. I have shown in Table 3 the number of birds in each age class for the five populations, each starting with 36 eggs.

Table 2. Stable age structure and demographic parameters of five hypothetical populations (A-E).

Age	A c_x	B c_x	C c_x	D c_x	E c_x
0	0.5461	0.5350	0.5503	0.6405	0.4834
1	0.1229	0.1408	0.1162	0.1441	0.1088
2	0.0897	0.0988	0.0862	0.0865	0.0859
3	0.0655	0.0693	0.0639	0.0519	0.0679
4	0.0478	0.0486	0.0474	0.0311	0.0536
5	0.0349	0.0341	0.0352	0.0187	0.0424
6	0.0255	0.0239	0.0261	0.0112	0.0335
7	0.0186	0.0168	0.0194	0.0067	0.0264
8	0.0136	0.0118	0.0144	0.0040	0.0209
9	0.0099	0.0083	0.0106	0.0024	0.0165
10	0.0072	0.0058	0.0079	0.0015	0.0130
11	0.0053	0.0041	0.0059	0.0009	0.0103
12	0.0039	0.0029	0.0043	0.0005	0.0081
13	0.0028		0.0032		0.0064
14	0.0021		0.0024		0.0051
15	0.0015		0.0018		0.0040
16	0.0011		0.0013		0.0032
17	0.0080		0.0010		0.0025
18	0.0006		0.0007		0.0020
19	0.0004		0.0005		0.0016

20			0.0004		0.0012
21			0.0003		0.0010
22			0.0002		0.0008
23			0.0002		0.0006
24			0.0001		0.0005
25			0.0001		0.0004
Total	1.0000	1.0000	1.0000	1.0000	1.0000
s_0	0.225	0.225	0.225	0.225	0.225
$s_{x>0}$	0.73	0.60	0.79	0.60	0.79
m_x	1.65	1.65	1.65	1.78	1.502
α	2	2	2	1	3
R_0	1.000	0.5549	1.3917	1.0000	1.0000
T	4.641	3.4599	5.6778	2.4738	6.6598
r	0.0000	-0.1564	0.0631	-0.0310	0.0000
E_0	1.6496	2.9738	2.0685	1.5613	0.8343

The three hypothetical populations (A, B, and C) with the same age-specific survival rates and fecundity are, again, barely distinguishable despite having different maximum ages. The two populations (D and E) that were constructed to have the same probability of survival (l_ω = 0.0008) of reaching maximum age as in Population A are quite different from Population A. The shorter-lived Population D is considerably smaller, and the longer-lived Population E is substantially larger, even though each started with 36 eggs and had l_ω = 0.0008.

Based on the results shown in Tables 2 and 3, we are probably justified in making the following generalizations: when individuals have short life expectancy, their populations' demography tends to show lower annual survival, greater fecundity, and earlier age of first breeding (think of ducks), and when individuals have longer life expectancy, their populations' demography tends to show greater annual survival, smaller fecundity, and later age of first breeding (think of albatrosses), although there is great variation.

With regard to the Ivory-billed Woodpecker's demography, I suspect that it is closest to that of Population A because other large woodpeckers live longer than 10 years but probably less than 20 years (Fitzpatrick et al. 2005), and the inclusion of non-breeding yearlings increases the likelihood of observing yearlings and families foraging and traveling together (Jackson 2004).

It is most important to note that the population structures given by the life history tables (Table 2) are for *large* persisting populations. The situation in *small* populations is quite different. In any population, the probability of any individual reaching maximum age is small (I assumed 0.0008). With l_ω = 0.0008 only three of 100 cohorts in populations A, D, and E may have an individual reach maximum age. Furthermore, in any population (Table

3), any age class represented by < 1 individual is not present in every cohort and, therefore, not present every year. The age classes in a cohort, which are represented by < 1 in Table 3, are usually represented in the corresponding real population by one individual, and *that* individual represents all of the subsequent age classes of that cohort. For example, in each cohort of Population A age 8 is expected to comprise, on average, 0.89 individuals, that is, one individual in some years, none in other years (Table 3). Regardless of the probability of survival, this individual either lives or dies. If it lives, then the cohort will have survived to at least age 9. If that individual dies, however, the cohort also dies at age 8, and subsequent age classes (9 to maximum age) in that cohort do not exist. Thus, cohorts at any time may be smaller or larger than the mean size expected from life table statistics, and populations at any time may not have all age classes represented (there may be 10-year-olds, but neither 9- nor 11-year olds)

In small populations, each population should be smaller than expected from the species' demography. If we discount the one "old" bird that may survive in a cohort (representing the cohorts numbering < 1 at each age in Table 3), the number of breeding adults of persisting Population A (represented by bold type) might be expected to be 18.6, of Population D, 18.7, and of Population E, 19.4 (Table 3). This corresponds to nine to ten pairs (if the sex ratio were 50:50).

With smaller $\overline{N}$, there will be a smaller $\overline{n}_0$, which results in a smaller $\overline{N}$, which results in a smaller $\overline{n}_0$, and so on to extinction (Table 3). Nevertheless, as populations get smaller, factors limiting survival or fecundity or both in larger populations may become less intense, resulting in a greater survival and fecundity and continued existence, albeit at a small population size.

On top of these purely demographic considerations is environmental variation. There may be good years and bad years for survival and breeding. A series of bad years spells disaster for a small population. But what happens in a good year following a bad year? If we increase m_x by 0.5 egg in their respective life tables, the per capita growth rate (r) increases from zero to 0.06025 in Population A, to 0.1072 in Population D, and to 0.04533 in Population E. Suppose instead they have a good breeding year with less mortality of young. If we increase s_0 by 50 percent, to 0.338, in their respective life tables, the per capita growth rate increases from zero to 0.0655 in Population A, to 0.18415 in Population D, and to 0.0822 in Population E.

DISCUSSION

Although we can never know the demography of the Ivory-billed Woodpecker, I suggest that its life table would resemble that of Population A (Table 1). Assuming that life history traits evolve when populations are relatively common (before anthropogenic interference), a persisting population with short life expectancy would have to evolve a larger clutch size and earlier age of first breeding, reducing the probability of observing nonbreeding yearlings or family groups after the young have reached independence. A persisting population with great life expectancy would have to evolve a smaller clutch size and later age of first breeding. The occurrence of broods of four or five young would be unlikely.

Compare populations A, D, and E when n_0 is 9, 18, 27, 36, and 45 eggs (Table 4). In Table 4 I have shown the numerical age structure for Population A with age classes of size $n_x > 1$ in bold. I have done the same for Populations D and E, but I have provided only a summary in Table 4. With a small population size, the populations with different life histories (A, D, and E) but the same $\bar{n}_0$ have a similar number of breeders despite sizable differences in $\bar{N}$. The mean population size of longer-lived individuals must be larger than those of shorter-lived species, if they are to sustain the same number of breeders and produce about the same number of eggs. (This is the demographic reason why such things as the Loch Ness Monster and Sasquatch do not exist. Being large they would have high survival and, unless they are immortal, would have to be part of a large population.)

Investigators may differ in their choice of the minimum sustainable population size. As I indicated above, my choice for a life history of the Ivory-billed Woodpecker would be closer to population A than to Populations D and E. If so, a population including ten breeding pairs, would number about 30 post-juvenile birds (I am now rounding off the numbers from those given in the analysis). With a longer life expectancy, the population with ten breeding pairs would number about 35 post-juvenile birds. There are several reasons for thinking that a persisting population could not be much smaller than 30 post-juvenile birds. First, the analysis assumes that the sex ratio is maintained close to 50:50. . If it is skewed very much, reproduction would certainly decline (I assume that polygamy is an unlikely possibility). A persisting population would have to be large enough to survive short-term periods of skewed sex ratios.

Table 3. Numerical age structure of five hypothetical populations with the proportional age structures given in Table 2. In this table, each population begins with $n_0 = 36$. In a small population with $n_0 = 36$, many age classes are represented by a fraction of one individual, which is of course not possible in life. The bold print refers to populations in which each age class numbers more than one. n_0^a = Breeders $\times$ m_x. $\mathbf{\mathit{n}_0^b}$ = **Breeders** $\times$ m_x. Age of first breeding is indicated by asterisk.

Age	A	B	C	D	E
0	36.000	36.000	36.000	36.000	36.000
1	**8.100**	**8.097**	**8.094**	***8.099**	**8.100**
2	***5.912**	***5.908**	***5.905**	**4.859**	**6.399**
3	**4.316**	**4.311**	**4.308**	**2.915**	***5.055**
4	**3.150**	**3.146**	**3.142**	**1.749**	**3.994**
5	**2.300**	**2.296**	**2.292**	**1.049**	**3.155**
6	**1.679**	**1.675**	**1.672**	0.630	**2.492**
7	**1.225**	**1.222**	**1.220**	0.378	**1.969**
8	0.894	0.892	0.890	0.227	**1.556**
9	0.653	0.651	0.649	0.136	**1.229**
10	0.477	0.475	0.474	0.082	0.971
11	0.348	0.347	0.345	0.049	0.767
12	0.254	0.253	0.252	0.029	0.606
13	0.185	0.000	0.184	0.000	0.479
14	0.135		0.134		0.378
15	0.099		0.098		0.299
16	0.072		0.071		0.236
17	0.053		0.052		0.186
18	0.038		0.038		0.147
19	0.028		0.028		0.116
20	0.000		0.020		0.092
21			0.015		0.073
22			0.011		0.057
23			0.008		0.045
24			0.006		0.036
25			0.004		0.028
			0.000		0.000
	29.919	29.273	29.912	20.201	38.465
Breeders	21.819	21.176	21.818	20.201	23.966
$\overline{N}$	**26.682**	**26.656**	**26.634**	**18.672**	**33.949**
Breeders	**18.582**	**18.559**	**18.539**	**18.672**	**19.450**
n_0[a]	36.001	34.941	35.999	35.999	35.997
	30.661	30.622	30.590	33.273	29.213

Second, a smaller population would be at greater risk from accidental extermination from a couple of bad years. Indeed, 30 birds may not be enough to sustain a persisting population through several poor years of survival and reproduction.

Nevertheless, some other populations had reached smaller sizes before serious conservation measures were implemented: the Whooping Crane (*Grus americana*) with 15-16 birds (Lewis 1995) and the Seychelles Warbler (*Acrocephalus seychellensis*) with 26 birds (Komdeur 1994). Neither species, however, persisted in that condition for fifty years. The possibility that a population of Ivory-billed Woodpeckers much smaller than 30 individuals could persist for fifty years seems unlikely.

Tanner (1942) gives three estimates of the density of Ivory-billed Woodpeckers: seven pairs in 120 square miles (one pair per 4,403 hectares) in the Singer Tract, six pairs in sixty square miles (one pair per 2,590 hectares) in the California swamp in Florida, and twelve pairs in 75 square miles (one pair per 1,619 hectares) in the Wacissa swamps in Florida.

Table 4. Numerical age structure of Populations A, D, and E. In this table the full numerical age structure is given for five populations of Population A, each starting with a different n_0. The bold print refers to age classes that number greater than one. $n_0 = $ **Breeders** $\times m_x$. Age of first breeding is indicated by asterisk.

Population A

Age	n_x	n_x	n_x	n_x	n_x
0	9.000	18.000	27.000	36.000	45.000
1	**2.025**	**4.050**	**6.075**	***8.100**	**10.124**
2	***1.478**	***2.956**	***4.434**	**5.912**	**7.391**
3	**1.079**	**2.158**	**3.237**	**4.316**	***5.395**
4	0.788	**1.575**	**2.363**	**3.150**	**3.938**
5	0.575	**1.150**	**1.725**	**2.300**	**2.875**
6	0.420	0.839	**1.259**	**1.679**	**2.098**
7	0.306	0.613	0.919	**1.225**	**1.532**
8	0.224	0.447	0.671	0.894	**1.118**
9	0.163	0.326	0.490	0.653	0.816
10	0.119	0.238	0.357	0.477	0.596
11	0.087	0.174	0.261	0.348	0.435
12	0.063	0.127	0.190	0.254	0.317
13	0.046	0.093	0.139	0.185	0.232
14	0.034	0.068	0.101	0.135	0.169
15	0.025	0.049	0.074	0.099	0.123
16	0.018	0.036	0.054	0.072	0.090
17	0.013	0.026	0.039	0.053	0.066
18	0.010	0.019	0.029	0.038	0.048
19	0.007	0.014	0.021	0.028	0.035
20	0.000	0.000	0.000	0.000	0.000
$\overline{N}$	7.480	14.959	22.439	29.919	37.398
Breeders	5.455	10.910	16.364	21.819	27.274
$\overline{N}$	**4.582**	**11.889**	**19.092**	**26.682**	**34.471**

Breeders	**2.557**	**7.839**	**13.018**	**18.582**	**24.346**
n_0	**4.22**	**12.93**	**21.48**	**30.66**	**40.17**

Population D—Summary

$\overline{N}$	5.050	10.101	15.151	20.201	25.252
Breeders	5.050	10.101	15.151	20.201	25.252
$\overline{N}$	**3.240**	**7.937**	**13.217**	**18.672**	**23.340**
Breeders	**3.240**	**7.937**	**13.217**	**18.672**	**23.340**
n_0	**5.773**	**14.143**	**23.553**	**33.273**	**41.592**

Population E—Summary

$\overline{N}$	9.896	19.792	29.688	39.585	49.481
Breeders	5.992	11.983	17.975	23.966	29.958
$\overline{N}$	**6.272**	**15.413**	**25.729**	**35.491**	**45.517**
Breeders	**2.367**	**7.604**	**14.015**	**19.872**	**25.994**
n_0	**3.555**	**11.421**	**21.052**	**29.850**	**39.046**

If these areas were isolated from other areas with Ivory-billed Woodpeckers, then these populations seem too small to be sustainable or at the lowest limit of sustainability. Could these be underestimates? Except for the first estimate, they are barely more than guesses. In the Florida swamps, the density was based on the assumption that A. T. Wayne and his collectors had killed "almost" all of the Ivory-billed Woodpeckers in each area. In the California swamp, Wayne had taken ten Ivory-billed Woodpeckers in two years. Tanner arbitrarily assumed that Wayne had missed a pair and presumed the population previously numbered six pairs. In the Wacissa swamps, Wayne had taken 19 Ivory-billed Woodpeckers in 1894. Tanner presumed that Wayne had not gotten them all (because a "few" persisted until 1937), and he estimated 12 pairs for the region prior to 1894. In the Singer Tract, Tanner (1942) was studying unmarked birds. Inasmuch as Ivory-billed Woodpeckers travel extensively and are not apparently aggressive (Tanner 1942), the possibility exists that he underestimated the population in the Singer Tract. I suggest that the density, much less the maximum density, of Ivory-billed Woodpeckers is best considered a guess.

Nevertheless, a small population, say 30 post-juvenile birds, could probably be accommodated within the 220,000 hectares (Fitzpatrick et al. 2005) of the Big Woods area in Arkansas (one pair per 15,000 hectares). Not all of those 220,000 hectares, of course, are suitable. If we generously accept Tanner's estimates of density (a mean of one pair per 2,870 hectares), then 15 pairs would require only 43,060 hectares of forest.

The issue remains, could a population as small as 15 pairs of Ivory-billed Woodpeckers have remained undetected in the Big Woods for more than 50 years?

ACKNOWLEDGMENTS

I thank J. W. Fitzpatrick, J. A. Jackson, and J. R. Jehl, Jr., for their commentary on earlier drafts of this ms. I thank K. Klimkiewicz for giving me the maximum age of the Pileated Woodpecker.

Literature Cited

ALLEN, A. A., AND P. P. KELLOGG. 1937. Recent observations on the Ivory-billed Woodpecker. Auk 54:164-184.

BENDIRE, C. 1895. Life histories of North American birds. Special Bull. No. 3, U. S. Natl. Mus.:42-45. (Not seen by me.).

BENT, A. C. 1939. Life histories of North American woodpeckers. U. S. National Museum Bull. 174.

BIRCH, L. C. 1948. The intrinsic rate of natural increase of an insect population. Journal of Animal Ecology 17:15-26.

Einstein, A., and Infeld, L. 1938. The evolution of physics explained by its most brilliant teacher. Addison-Wesley.

FITZPATRICK, J. W., M. LAMMERTINK, M. D. LUNEAU JR., T. W. GALLAGHER, B. R. HARRISON, G. M. SPARLING, K. V. ROSENBERG ET AL. 2005. Ivory-billed Woodpecker (*Campephilus principalis*) persists in continental North America. Science 308:1460-1462.

FITZPATRICK, J. W., M. LAMMERTINK, M. D. LUNEAU, JR., , T. W. GALLAGHER, AND K. V. ROSENBERG. 2006. Response to Comment on "Ivory-billed Woodpecker (*Campephilus principalis*) persists in contental North America." Science 311.

JACKSON, J. A. 2002. Ivory-billed Woodpecker (*Campephilus principalis*) *in* The Birds of North America, vol 711 (A. Poole and F. Gill, eds.). The Birds of North America, Inc., Philadelphia, PA.
JACKSON, J. A. 2004. In Search of the Ivory-billed Woodpecker. Smithsonian Books, Washington, D. C.

JACKSON, J. A. 2006. Ivory-billed Woodpecker (*Campephilus principalis*): hope, and the interfaces of science, conservation, and politics. Auk 123:1-15.

KOMDEUR, J. 1994. Conserving the Seychelles Warbler *Acrocephalus sechellensis* by translocation from Cousin Island to the islands of Aride and Cousine. Biological Conservation 67:143-152.

LEWIS, J. C. 1995. Whooping Crane (*Grus americana*) *in* The Birds of North America, vol. 153 (A. Poole and F. Gill, eds.). The Academy of Natural Sciences, Philadelphia, PA, and The American Ornithologists' Union, Washington, D.C.

LOTKA, A. J. 1925. Elements of Physical Biology. Williams & Wilkins, Baltimore.

MURRAY, B. G., JR. 2003. A new equation relating population size and demographic parameters: some ecological implications. Annales Zoologici Fennici 40:465-472.

MURRAY, B. G., JR., J. W. FITZPATRICK, AND G. E. WOOLFENDEN. 1989. The evolution of clutch size. II. A test of the Murray-Nolan equation. Evolution 43:1706-1711.

MURRAY, B. G., JR., AND L. GÅRDING. 1984. On the meaning of parameter x of Lotka's discrete equations. Oikos 42:323-326.

MURRAY, B. G., JR., AND V. NOLAN, JR. 1989. The evolution of clutch size. I. An equation for predicting clutch size. Evolution 43:1699-1705.

NOLAN, V., JR. 1978. The Ecology and Behavior of the Prairie Warbler *Dendroica discolor*. Allen Press, Lawrence, KS.

SIBLEY, D. A., L. R. BEVIER, M. A. PATTEN, AND C. S. ELPHICK. 2006. Comment on "Ivory-billed Woodpecker (*Campephilus principalis*) persists in continental North America." Science 311.

TANNER, J. T. 1941. Three years with the Ivory-billed Woodpecker, America's rarest bird. Audubon Magazine 43:5-14. (Not seen by me.).

TANNER, J. T. 1942. The Ivory-billed Woodpecker, National Audubon Society.

WOOLFENDEN, G. E., AND J. W. FITZPATRICK. 1984. The Florida Scrub Jay: Demography of a Cooperative-breeding Bird. Princeton University Press, Princeton.

Commentary (*Auk*, 6 July 2006)

[Ornithologists have very little experience with life history tables, as attested by the fact that virtually no life-history tables have been published in the past 50 years. Papers on "demography" are numerous but treatment is cursory, usually focusing on one or two variables, such as age-specific fecundity or annual survival rates. Life-history tables can tell us much more about populations. In addition to the net reproductive rate, we can calculate generation time, life expectancy at birth, population growth rate (r), and so on, about which we ornithologists know little. Furthermore, papers on demography are descriptive rather than analytical. Thus, a paper, such as mine on the Ivory-billed Woodpecker, is a big surprise to them. How is it possible to create a life-history table for an apparently extinct species about which little is known? Well, of course, that is the point. Theory not only explains that which we think we know, but it should also inform us about things that we cannot otherwise know. A study of life-history tables shows that there is a limited number of allowable combinations of life-history traits (Chapter 3). Some combinations are just not possible. Thus, if you tell me that a species has a large clutch size, I will tell you that it has a poor survival rate. I do not tell you this because of empirical studies that show the correlation but because of a *mathematical theory* that shows that relationship. A theory explains what we see; it is *not* derived from correlations that we have observed in nature. (As Einstein and Infeld [1938:8] pointed out about Newton's First Law, "We have seen that this law of inertia cannot be derived directly from experiment, but only by speculative thinking consistent with observation. The idealized experiment can never be actually performed, although it leads to a profound understanding of real experiments.") Sooner or later, theory must escape from the textbooks and be used to infer facts.

This paper describes how to create a life history table for an apparently extinct species. I propose what I think is a likely life-history table for the Ivory-billed Woodpecker. Certainly, the Ivory-billed Woodpecker has, or had at one time, a life history. Those who disagree with me should explain why a theoretical treatment of the life history of the Ivory-billed Woodpecker should be uninteresting. If they think that it might be interesting, then perhaps they should propose an alternative life history that seems more likely (at least to them).

Although my paper is ostensibly about the Ivory-billed Woodpecker, it is also an exercise on the dynamics of small populations, which is not quite the same as that of large populations, a matter that might be of some interest to those who study small populations. I think the paper is important in showing a method of analysis of populations, which has never been done before, as far as I know. I presume that other ecologists might be interested in such problems. There is a huge literature, and I certainly have not seen everything. Perhaps this kind of analysis has been done by someone else, but I do not know of it. If any reader should find one, I should appreciate being apprised of it.]

LETTER TO THE EDITOR

Dear …:

It is no surprise to me that the reviewers found my ms. (06-081), "Demography and population dynamics of the Ivory-billed Woodpecker: the dynamics of small populations," not worthy of publication. Reviewers seem not to like anything that I do. I think the reason is that I am in entirely different philosophical and scientific paradigms from so many of my colleagues.

Because I really have no way of communicating with my colleagues through the journal literature, I write commentaries on the reviews, with the hope that editors will forward them on to the reviewers, and with the hope that the reviewers may come to understand why I disagree with them on so many things.

First, when I got interested in birds, I also got interested in the lives of birds and their evolution. I can remember the moment when the scales fell from my eyes. I was reading a paper by David Lack, who was explaining the migration of Redwings from Iceland to England. In the course of the paper, it became clear to me that he was proposing a different explanation from that of Kenneth Williamson. What a shock to me. Everything was not already known about biology (as it had seemed to me from my college courses—according to which *everything* was already known). So, my life had direction. I was going to find out more about the lives of birds, and this is what I have done. As things turned out, I eventually focused on population dynamics. And, as things turned out, we ornithologists/ecologists/evolutionary biologists no longer discuss the pros and cons of alternative explanations. We no longer have debates like we had between Lack and Williamson and between Lowry and Williams. In other words, we no longer debate each other because only one side of the argument can be heard. We do not debate any more because one side, the one in control of what gets published, believes that it already knows everything.

Second, philosophically, I think that whatever I think is an explanation of some fact or process must at some point be supported by evidence and argument. I am not allowed to think whatever I want to think. Indeed, as Feynman has said (paraphrasing), "Everything I know must be consistent with everything else I know."

For example, in summarizing the reviewers' comments, you say, "As a deterministic model, it simplifies a lot of population ecology that would be relevant to population viability, including density-dependence, a stochastic environment, metapopulation dynamics, and genetic considerations."

It seems unknown among biologists that I have been challenging the idea of density-dependent regulation of population numbers since 1979 (Murray 1979, 1982, 1994, 1999a, 1999b, 2000a, b). I have asked, several times, "What is the density-dependent regulation hypothesis?" I know what it used to be, prior to 1980, but I do not know what it is now, and no one will tell me.

(1) I proposed a model (Murray 1979, 1982) in which I kept age-specific survival and the fecundity of breeding females *constant at all densities*. Yet, the population stopped growing. No one has thought to show any flaws in my model, much less in my thinking. Indeed, no one even

acknowledges its existence (never been cited). Now that certainly is an example of a scientific debate about a central issue in ecology[3].

I constructed age-structured populations from a random numbers table that persisted for up to 100 years (Murray 2000a). These populations, of course, lacked density-dependent negative-feedback loops. The reviewers recommended acceptance of my paper if they had the opportunity to publish a commentary. The criticisms in their commentary were based on misunderstandings (Murray 2000b).

More recently, I showed that a population of Pied Flycatchers that grew from one pair in 1962 to 125 pairs in 1990 did not grow logistically. Yet, Saether et al. (2002) "fit" a logistic curve to the data.[4]

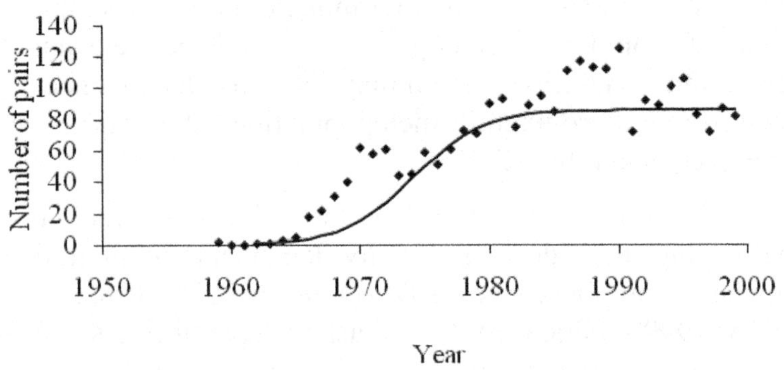

---

[3] I hope that those readers who do not know me recognize my sarcasm here.

[4] They did not actually plot their theta-logistic curve on their graph of population numbers, as I have done here. I wonder if they thought everyone else would also notice the extremely poor fit.

I think this is a pretty poor fit. Instead, I "fit" a linear regression to the data from 1962 through 1990.

The Pied Flycatcher population is not growing logistically because (i) the curve is not S-shaped between near-zero and maximum population size, (ii) the plot of ΔN vs. N is not dome-shaped, and (iii), at the least, ΔN does not decrease above $0.5\ K$ with increasing density. All these are characteristics of logistic growth. Nevertheless, I am unable to get my paper on this published anywhere (rejected from five or six journals). So much for dialogue on important ecological issues.

So, I ask again, What is the evidence for density-dependent regulation? I contend that there is none whatsoever. I make such a sweeping statement in order to make its refutation easy on my critics. All they have to do is find one good case to refute that statement. No ecologists have written to me at all, much less given me what they think is their best evidence.

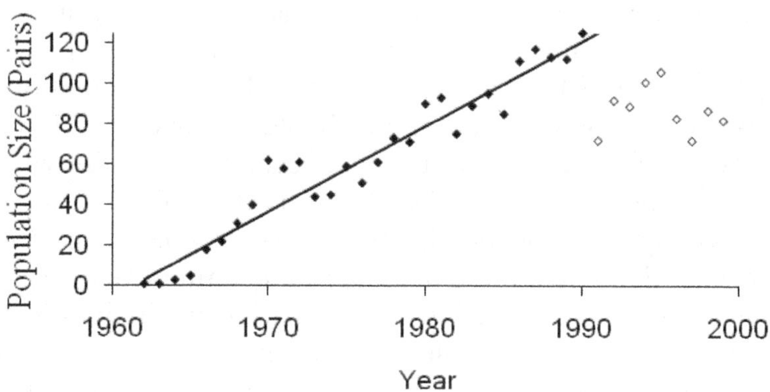

The reason that I do not discuss density-dependent regulation in my paper is because it is a figment of the ecologists' imagination. Furthermore, it is irrelevant to the problem I was studying.

(2) "Determinism" is a word that biologists seem not to understand. They *always* use it as a pejorative. Determinism in the sense of "biological determinism," which is a *philosophy* that uses biology (e.g., genetics) as a justification for social injustice (racism, sexism, nationalism, etc.) around the world, is reprehensible (which itself is also a philosophical position).

The word determinism, however, has other uses. For example, every mathematical equation is deterministic. If a = 3, and if b = 5, then a + b = 8; 3 + 5 *always* equals 8. How much more deterministic could determinism be? Nevertheless, a + b does not *always* equal 8. The reason is the "if" clauses. We cannot tell how many pieces of fruit are in a basket unless we know there are three apples and five peaches. This is true of all mathematical equations; all mathematical equations are deterministic, but no mathematical equation tells us what the world should be like. Would you want it any other way?

In the same way, all theories (well, all testable theories) are deterministic. Theories take the form, *if* a, *if* b, *if* c, *then* d (otherwise known as *explanans* (a, b, c) and *explanandum* (d)). From his laws and initial conditions, Newton predicted that an apple should fall about 16.1 feet (he was more accurate) in the first second of free-fall when near the earth's surface. The theory is deterministic (given the assumptions, the prediction cannot be other than 16.1 feet), but the world itself need not be deterministic. That is why we must test theories by comparing predictions with empirical data. We would actually have to go out and measure how far an apple falls in one second near the earth's surface. It would not *have* to fall 16.1 feet because the real world is not deterministic and the theory could be wrong. Think about it, theories could not be predictive without also being deterministic. What is a prediction if not a determinist conclusion from a set of assumptions? Theories could not be tested if they were not deterministic. How else could science work? (I recommend that biologists read Popper (*The Open Universe: An Argument for Indeterminism*).[5]

(3)   With regard to both points (1) and (2), one presubmission reviewer (his name is not included in the Acknowledgments) wrote, "the most important parameter, which is the variance in the growth rate due to environmental and demographic stochasticity, is not estimated nor mentioned." I wrote to him (we are corresponding, reluctantly on his part) and asked how he would go about calculating the variance of the growth rate of a near extinct (if not already extinct) species. He replied, "In regards to the variance on r. You are right [Wow! Did I read that correctly?]; we do not have data about it for the woodpeckers (in fact for precious few populations of anything). However theory shows that at low densities and for low values of r its variance Var(r) determines the fate of the population." Wow! Did I read that correctly? Var(r) *determines* the fate of the population? He criticizes me for being deterministic ("Dr. Murray's approach is entirely deterministic."), and then turns around and tells me that some theoretical statistical property of theoretical growth rates *determines* the outcome of a biological process.

I asked this person for empirical evidence that Var(r) was responsible for the extinction of *any* population. I thought this might be challenging, considering that he had already admitted that Var(r) is known for precious few species of anything, and he provided no references to me for those few. Why do ecologists believe this stuff? Just because it appears in some textbooks? And it *looks* sexy? I would like to see (at least) a comparison of Var(r) values for extinct and extant populations. None exists, of course. For me, I dismiss

---

[5] Sic.

this kind of stuff because it is speculation without any body of evidence. I don't even try to remember it.[6]

Specifics

(1) *I want you to note that the reviewers did not identify a single sentence, fact, or calculation as incorrect.* The first criticizes me for putting in too much. The second criticizes me for not putting in enough. The reviewers' comments are essentially philosophical—how we should go about doing science and writing papers.

(2) The first reviewer suggests, "I think the MS could be less pedantic (this would amount to toning down Murray's inimitable style a bit) and perhaps restrict itself to fewer options, both of which would allow the paper to be shortened without compromising the conclusions. Do we really need all the various life-table calculations? Perhaps once (Table 1) but I doubt we need cx values or age structure for 5 hypothetical populations (Table 2, 3, and 4)." Well, I contend that I am not a pedant, although I consider myself a teacher. I just do not know what my "inimitable" style is, unless it is explaining what I think in as clear language as I can.

The reason that I included an eight-column life table is because these do not appear anywhere in the ecological literature (other than my papers), and therefore I assume that my readers are unfamiliar with them. "Anywhere in the ecological literature" includes ecological textbooks. I have not seen a textbook with an instructive life table from which putative ecologists (i.e., students) could learn what they are about, much less how to construct and interpret one. I had to learn myself, and I am still learning after having done

---

[6] This kind of stuff actually clutters the mind. Perhaps, this is why Nobel Prize winner Albert Szent-Gyorgyi stated that he paid little attention to the literature.

thousands of them during the past 30 years. From my reading of the ornithological/ecological/evolutionary literature, I am not convinced that my colleagues are knowledgeable about the relationships of demographic parameters (just for example, consider those ornithologists who still believe that greater predation on eggs or nestlings selects for smaller clutch size, a mathematical impossibility).[7] I am sure they are comfortable with their knowledge, but that is only because they do not demand that their theoretical knowledge correspond (even once in a while) with the empirical facts.

The reason that I included the age structures of five populations (Table 2, 3, and 4) is simply that, again, I just do not believe that ecologists understand age structure sufficiently well to juggle the ideas in their heads. After all, the age structure of populations is not commonly (if at all—I don't know of any) discussed in the literature. (Furthermore, despite my very explicit correction [(Murray 1979)] and brief statements by others, textbook authors [and of course their readers] continue to misinterpret age pyramids. The broad-based pyramids of places like India are not indicators of high birth rates in those populations but of high mortality rates in the younger age classes.[8]) The tables in my paper provide a visual image that (I believe) just does not come across in words alone. A reader can actually *see* the age structure rather than *imagine* the meaning of my verbal comparisons and conclusions. I think that ecologists would think a lot differently about things if they actually worked through the mathematics of life tables and other problems. The reviewer may think that I could cut these out "without compromising the conclusions," but I do not.

The reviewer states, "After all, as Murray points out, with such a small (likely) population to begin with, those 4 decimal place proportions are pretty much irrelevant." In this

---

[7] I pointed this out first in 1979.

[8] I pointed this out first in 1979, but I would not claim priority.

regard, I wrote in the ms., "In working with life history tables, it is best to 'estimate' numerical parameters to three, four, or five decimal places so that small but important numerical differences do not disappear and because there are several ways to cross-check the calculations precisely. We can always round off our final numbers." If the reviewer, as one reader, does not like the numbers to four decimal places, he can easily cross out the two[9] extra decimals. Another reader, who might want to see the numbers to three or four decimals could not easily reconstruct them if I left them out. The reviewer should remember that in "deterministic" theoretical work, I can write out as many decimals as I wish. It is in empirical work, in which the investigator is working with sample sizes and sampling error that one cannot write out an arbitrary number of decimal places.

In the Discussion I wrote, "(I am now rounding off the numbers from those given in the analysis)."

Ecologists give the impression of being more interested in the quality of the trim of their car than in the quality of the engine.

(3) Both reviewers are concerned with the lack of "hard" data. The second, especially, writes, "very little is known about the demographic traits of this species. Basically, we are dealing with anecdotes and an enormous level of uncertainty for every assumption and specific quantity that is plugged into what are very basic equations." I wonder what the reviewer can possibly mean. Does he think that the clutch size of the Ivory-billed Woodpecker might be ten eggs or fifteen, or maybe two or four (I assumed three)? I will work out a life table for him for any assumptions that he wishes to make. Let's see what happens. Of course, remember, the first reviewer thought that extra life tables were unnecessary. I picked a modal clutch size of three eggs because I thought

---

[9] Or three?

that that was most likely from the *real* data that were available.

Does the reviewer think that Ivory-billed Woodpeckers could live for 20, 30, or 40 years, as does Jerry Jackson? I don't know; he does not say. Actually, I picked a number between 15 and 20 because Lammertink, an apparent expert on large woodpeckers, says that large woodpeckers rarely (if ever) live longer than 15 years (I believe him), and, as a nod to Jerry, who believes, on the basis of the 17 years that his pet Red-cockaded Woodpecker lived, that Ivory-bills lived longer than 15 years. Furthermore, as I pointed out, the maximum potential age of the Ivory-billed Woodpecker or any species is irrelevant when we are concerned with the minimum likely size of a *persistent* small (less than thousands) population.

Is the reviewer correct that we can say *nothing* of interest about the demography of the Ivory-billed Woodpecker (even theoretically)?

(4) The reviewer states, "One problem with this paper is that much (certainly through page 8) of it [sic] devoted to very generic/basic population ecology that has little really little [sic] to do with the IBWO. Things improve a bit on page 7 with the commentary on small population, but – even there – the approach is classic/deterministic with no recognition that small populations may have important intrinsic differences (due to, say, inbreeding) and other problems owing to influences such as Allee effects. These issues may, indeed, be irrelevant but Murray proceeds as if persistent small population size is important only to age distribution and then actually speculates that release from negative density dependence (a force that is not in any of the models) will have a positive effect on survival and fecundity."

Where to begin? Actually, as I indicated above, I believe that ecologists need to state their demographic assumptions in their papers because there is no agreed upon "basic population ecology" (I say this not just because I disagree with what has been written in textbooks and papers on demography and population dynamics, but because the authors of textbooks and papers are not themselves consistent, and they disagree with each other). Furthermore, my analysis and conclusions are based mostly on Equation 1, which does not appear in any ecological publication other than my own (Murray 2003). Although I think Equation 1 should be part of basic population ecology, it is not. My introductory material in Population Equations and Hypothetical Life History Tables constitutes the Methods portion of my paper. Shouldn't I explain my methods? Are those sections not recognizable as methods even if not labeled so? Inasmuch as readers will not know of Equation 1 and none works with life tables (at least I have not seen a published life history table, such as my Table 1, for any bird species in the past 45 years), I think it is necessary for me to explain my methods. Can you imagine the hullabaloo that would occur if I did not?

The reviewer criticizes me for not discussing issues that he says "may, indeed, be irrelevant." Good heavens! I find this quite exasperating.[10]

The reviewer states that Murray "actually speculates that release from negative density dependence (a force that is not in any of the models) will have a positive effect on survival and fecundity." I certainly never speculated about negative density dependence because, as I indicated above, I am the chief critic of density-dependent regulation of any kind. I think the reviewer should read what I think about population dynamics (start with Murray 1979), which is the context in which I write. I write in my context (paradigm).

---

[10] Although exasperating to me, editors seem to buy this nonsense. "Bert Murray has left out discussions of irrelevant material. By all means, reject!"

The reviewer reads in *his* context (paradigm). Why should his context, without evidence or arguments, prevail?[11]

(5) The reviewer states, "Issues of density, spatial structure, the spatial scale of ?a population? [sic; I have no idea what these question marks should be. It appears they should be quote marks but quote marks appear elsewhere in the review (e.g., see the review's last sentence.)] in this species, and demographic-environmental stochasticity are essentially ignored."

This is puzzling. I wrote, for example, "On top of these purely demographic considerations is environmental variation. There may be good years and bad years for survival and breeding. A series of bad years spells disaster for a small population. But what happens in a good year following a bad year? If we increase m_x by 0.5 egg in their respective life tables, the per capita growth rate (r) increases from zero to 0.06025 in Population A, to 0.1072 in Population D, and to 0.04533 in Population E. Suppose instead they have a good breeding year with less mortality of young. If we increase s_0 by 50 percent, to 0.338, in their respective life tables, the per capita growth rate increases from zero to 0.0655 in Population A, to 0.18415 in Population D, and to 0.0822 in Population E."

And, "First, the analysis assumes that the sex ratio is maintained close to 50:50. If it is skewed very much, reproduction would certainly decline (I assume that polygamy is an unlikely possibility). A persisting population would have to be large enough to survive short-term periods of skewed sex ratios. Second, a smaller population would be at greater risk from accidental extermination from a couple of bad years. Indeed, 30 birds may not be enough to sustain a

---

[11] Every statement that I have made during the past 45+ years has been accompanied by evidence and arguments. I may have been wrong, but I always have offered evidence and arguments. And no one has yet shown that I have been wrong.

persisting population through several poor years of survival and reproduction."

How am I "essentially" ignoring stochasticity? Have I ignored stochasticity because I have not put it in a deterministic equation?

Finally, on this, I believe it is necessary to understand the simple case, that is, the situation of a stable age distribution, before we can speculate on deviations. Trying to understand variations without understanding the simple case would seem to lead us into confusion. Casting physical problems into the simple cases was *explicitly* the method of Isaac Newton, Albert Einstein, and Richard Feynman. Why shouldn't we biologists learn something from the physicists?

(6) The reviewer states, "With all due respect, the result that there ?should be? [sic] about 15 pairs is so uncertain that it may do more harm than good. By any stretch, a population of 15 pairs will not persist. Will agencies and managers read this and conclude the the [sic] population is 'Doomed to extinction?'" I never suggested that the population of Ivory-billed Woodpeckers in the Big Woods "should be" about 15 pairs. What I suggested was that the population, if it were persisting, could probably not have a mean population size *of fewer than* 15 breeding pairs. In other words, 15 breeding pairs was a minimum mean over a period of time, not necessarily *the* true population size at any or every time.

If agencies and managers draw the conclusions the reviewer has them draw from my paper, then they are as deterministic in their thinking as the reviewer. Both reviewers read far more deterministically than I write.

In my cover letter to you, I wrote that after hearing of the discovery of the Ivory-billed Woodpecker (whether true or not—I saw the video and thought it was true), my first thought was about how many Ivory-billed Woodpeckers would have to be there. I just could not buy the concept that there is one very old lone bird flitting around in the swampy

South. So, I wanted to know what a persisting population would have to look like. In my paper, I explained in detail how I went about arriving at an answer. I explained it in detail because I thought the train of thought would not be immediately obvious to biologists who do not work with life history tables (and because nothing written about life tables—construction or interpretation—seems to be accurate). I also thought that my research might be of interest to more ornithologists than just to me.

Cheers,

Bert

Literature Cited

Murray, B. G., Jr. 1979. Population Dynamics: Alternative Models. Academic Press, New York.

Murray, B. G., Jr. 1982. On the meaning of density dependence. Oecologia 53:370-373.

Murray, B. G., Jr. 1994. On density dependence. Oikos 69:520-523.

Murray, B. G., Jr. 1999a. Can the population regulation controversy be buried and forgotten? Oikos 84:148-152.

Murray, B. G., Jr. 1999b. Is theoretical ecology a science? A reply to Turchin (1999). Oikos 87:594-600.

Murray, B. G., Jr. 2000a. Density dependence: reply to Tyre and Tenhumberg. Austral Ecology 25:308-310.

Murray, B. G., Jr. 2000b. Dynamics of an age-structured population drawn from a random numbers table. Austral Ecology 25:297-304.

Murray, B. G., Jr. 2003. A new equation relating population size and demographic parameters: some ecological implications. Annales Zoologici Fennici 40:465-472.

Saether, B.-E., S. Engen, R. Lande, C. Both, and M. E. Visser. 2002. Density dependence and stochastic variation in a newly established population of a small songbird. Oikos 99:331-337.

[I do not receive replies to my letters, so I am unable to determine why anyone should think that I should think that I am wrong.]

Chapter 5

Clutch Size and Length of Breeding Season

[The increase in clutch size of birds from low latitudes to high latitudes has been well documented. The reasons have been debated for over 60 years. There has been no resolution because ornithologists provide ad hoc untestable explanations from which no independent predictions can be made. I wrote a paper reviewing geographical variations in clutch size and showed that clutch size varied with the length of the breeding season. This paper was submitted in various forms under different titles to several journals without success. I first submitted it to the *Wilson Bulletin* ("On the evolution of large clutch size in tropical passerines") on 16 July 1993. It was rejected on 14 September 1993. I next submitted this paper to the *Auk* on 27 September 1993 but withdrew it on 25 January 1994, no doubt because I had no confidence that the editor would get the manuscript reviewed in anything like a reasonable period of time. As it turned out, my later patience was unrewarded. I submitted the ms. to the *Journal of Avian Biology* on 23 January 1994 (revised and resubmitted, 15 June 1994), and to the *Condor* on 7 March 1995 (revised and resubmitted, 3 July 1995). I changed the title of the paper to "Large clutch sizes in tropical passerines: tests of theory" and rolled the dice with the *Journal of Tropical Ecology*, submitting it on 14 March 1996. And I tried again, submitting it under another title, "The evolution of large clutch sizes in tropical passerines," to *Biotropica* on 12 April 1999.

I really believed that my paper was an important contribution to the evolution of clutch size in birds. So, I persisted. Stimulated by a paper by Eleanor Rowell and Ian

Rowley on clutch size, I changed the title of the paper to "Why the clutch sizes of passerine birds are smaller in the Southern Hemisphere than in the Northern Hemisphere," revised the emphasis, included more examples, and submitted it to *Austral Ecology* on 22 February 2000. Finally, I submitted this ms. to *Oikos* on 5 July 2001.

The manuscript sent to *Oikos* is reproduced here.]

MANUSCRIPT

Why the clutch sizes of passerine birds are smaller in the Southern Hemisphere than in the Northern Hemisphere

Abstract

The mean clutch size of birds varies inversely with the length of the breeding season. It is greater at higher latitudes than in the tropics, greater in the northern hemisphere than at comparable latitudes in the southern hemisphere, and greater in species with shorter breeding seasons at the same latitude. Although small clutch sizes and long breeding seasons have been noted in Australia and South America, no causal relationship, other than mine (Murray 1979, 1985, 1991a), seems to have been proposed. Clutch sizes should be smaller where breeding seasons are longer because females with longer breeding seasons should have a higher probability of rearing young. The clutch size of birds, thus, should be larger in species with shorter breeding seasons and smaller in species with longer breeding seasons. In this paper I review the evidence for the relationship and explain why such a relationship should be expected from theory.

Introduction—The Problem

Tropical passerines typically have clutches of two or three eggs, whereas passerines at higher latitudes in the northern hemisphere have four or more eggs (Lack 1968, Klomp 1970, Skutch 1985). Mean clutch size increases less steeply with increasing latitude in the southern hemisphere (Moreau 1944, Yom-Tov 1987, Rowley and Russell 1991, Yom-Tov et al. 1994). Lack (1947, 1948) suggested that the long day length during the breeding season at high latitudes allowed parents to find, capture, and deliver sufficient food for rearing larger broods than parents of tropical species could rear with their shorter day lengths. Skutch (1949) disagreed and proposed that small broods in the tropics were less subject to predation than large broods.

Clutch size in the southern hemisphere is smaller than in the northern hemisphere. The mean clutch size of 299 species of passerine birds in Australia is 2.7 eggs (Yom-Tov 1987, Rowley and Russell 1991), of 353 species in southern Africa, 2.8 eggs (Rowley and Russell 1991), and of 331 species in temperate South America, 2.98 eggs (Yom Tov et al. 1994). The southern continents do not extend as far south as the northern continents extend north, making the comparison inexact. Nevertheless, the clutches of the southern hemisphere are smaller than at the more comparable latitudes of northern Africa, about 4.3 eggs for 119 passerine species (Yom-Tov 1987, Rowley and Russell 1991).

In the southern hemisphere the increase in clutch size with latitude is smaller than in the northern hemisphere. In eastern Australia the mean clutch size increases with increasing latitude from 2.18 for 40 passerine species from the York Peninsula to 2.71 for 38 passerine species from New South Wales (Yom-Tov 1987). The populations of more than half the passerine species that range from tropical Africa to South Africa have a clutch a little less than one-half egg larger than tropical populations (Moreau 1944). In South

America, which extends much farther to the south than either Africa or Australia, no latitudinal increase in mean clutch size had apparently been noted by Johnson (1965) and Yom-Tov et al. (1994).

It should be understood that many of these estimates of mean clutch size from Australia and South America are means of means, the latter of which are too often estimates from ranges. For example, if a clutch size was reported as "2 eggs, sometimes 3," it was given a mean of 2.3; if clutch size was reported as "2 eggs, sometimes 1," it was scored as 1.7 (Yom-Tov 1987, Yom-Tov et al. 1994). Thus, the data should not be interpreted too rigorously. For example, Yom-Tov et al. (1994) did not find latitudinal variation in the mean of mean clutch sizes of the two suborders of passerines in southern South America, a conclusion reached earlier by Johnson (1965). Combining the clutch sizes of many species of a suborder could obscure existing variation within species. For example, Johnson's (1965:381) claim that "there is no evidence that clutch size tends to increase with latitude ... even where the same species nests both in the extreme south and in the semi-tropical north" seems contradicted by his later statement with regard to the rufous-collared sparrow *Zonotrichia capensis* in northernmost Chile, "as in Antafagasta [also in northern Chile], almost all nests had only 2 eggs in contrast to the usual 3, 4 or even 5 of the southern races" (Johnson 1967:392). Also, in the house wren *Troglodytes aedon,* the mean clutch size increases from 3.25 in Arica, Chile, to 5.00 in Tierra del Fuego (Young 1994a). Thus, despite the limitations of the data, the broad pattern, that clutch size within species increases with increasing latitude in the northern hemisphere and that clutch size increases to a lesser extent in the southern hemisphere, seems correct.

Yom-Tov et al. (1994) suggested that the smaller clutch size in the southern hemisphere was best explained by Ashmole's hypothesis, as described by Ricklefs (1980). The idea is that because northern climates during the non-breeding season are more severe than in the southern hemisphere, mortality is greater in northern hemisphere birds (because of the greater hazards of migration or because food was less available) than in the southern hemisphere where the climates are milder. Thus, at the beginning of the breeding season, northern populations are small relative to the newly available flush of resources in spring and summer and are able to rear large broods, whereas southern populations are large relative to available resources and cannot rear such large broods. In commenting on Ashmole's hypothesis, Hussell (1985:634) was correct when he observed that the "possibility remains that the latitudinal trend in clutch size may be explicable in terms of any of these variables [e.g., actual evapotranspiration (AE) in winter, AE in summer, daylength, and length of breeding season] or a complex relationship among several of them or others not considered here. ... Demonstration of a correlation between clutch size and an environmental variable provides little insight into the causative agents of clutch size determination and merely allows that variable to remain as a viable candidate for consideration as an element of an explanatory hypothesis. We need to find other ways to test such hypotheses."

Martin et al. (2000) pointed out, "The inability of the most widely invoked hypotheses to explain latitudinal patterns in clutch size illustrate [sic] that alternative hypotheses ... deserve more attention, and that current theories of clutch size evolution need major revision." Martin et al. (2000), however, did not consider my theory on the evolution of clutch size (Murray 1979, 1985, 1991a), which predicts the clutch-size variations just described.

My approach has been to eschew correlations because correlations always have exceptions, and those exceptions also need explanation. A *post hoc* explanation of a correlation does not often explain the exceptions. Instead, I predict clutch size and clutch-size variations from first principles (i.e., now called laws [Murray 1999, 2000a, 2001a]). From these (see below) I predicted that populations with long breeding seasons should have smaller clutches than populations with short breeding seasons (Murray 1979, 1985, 1991a). Thus, because of the small clutch sizes in the southern continents, one should predict that breeding seasons are longer in Australia, Africa, and South America than at comparable northern latitudes. This prediction seems confirmed (Baker 1938, Wyndham 1986, Yom-Tov et al. 1994, and Rowley and Russell 1991). My theory, then, seems to explain why clutch size is smaller in the tropics than at higher latitudes (longer breeding season) and smaller in the southern than in the northern hemisphere at comparable latitudes (again, longer breeding season).

Although others have noted that small clutch sizes occur in regions with longer breeding seasons (Rowley and Russell 1991, Yom-Tov et al. 1994), no one has recognized (despite my earlier predictions; Murray 1979, 1985, 1991a) a direct cause-and-effect relationship between them, simply because a causal connection is not intuitively obvious. The purpose of this paper is to show why this relationship between the length of the breeding season and clutch size is expected from the laws of my theory.

Theory

My theory begins with three universal laws (Murray 1999, 2000a, 2001a): (1) the genotypes and phenotypes with the greatest Malthusian parameter increase more rapidly than those with smaller Malthusian parameters; (2) in the absence of changes in selection forces, a population will reach and remain in an evolutionary steady-state; and (3) selection favors those females that lay as few eggs as are consistent with replacement because (i) they have the highest probability of surviving to breed again, (ii) their young have the highest probability of surviving to breed, or (iii) both. The logical and empirical justifications for these laws have been presented elsewhere (Murray 1979, 1985, 1991a, 1999, 2000a, 2001a).

From these assumptions, Murray and Nolan (1989) proposed an equation for calculating the mean clutch size (C) of females from other population parameters,

$$C = \frac{a+1}{\sum_{\alpha}^{\omega} \lambda_x \sum_{1}^{n} P_i},\qquad(1)$$

where a is the primary male/female sex ratio (assumed to be 1 in birds), λ_x is the probability of surviving from birth (in birds, from the laying of the egg) to age class x of those individuals from successful clutches or litters, α is the average age class of first breeding, ω is the age class of last breeding, and $\sum_{1}^{n} P_i$ is the mean number of broods reared during a breeding season. $\sum_{1}^{n} P_i = c_1 s_1 + c_2 s_2 + \ldots + c_n s_n$, where c_1, c_2, and c_n are the mean number of eggs laid in producing the first, second, and nth brood, and s_1, s_2, and s_n are the probabilities that a first, second, and nth brood clutch is successful in producing at least one young to independence (Murray 1991a, 1991b, 2000b).

According to Wootton et al. (1991), Eq. 1 must be true, and I agree with them (Murray 1992). This equation must hold, whether one accepts Lack's hypothesis, Skutch's hypothesis, my hypothesis, or some other hypothesis as the best explanation for the evolution of clutch size. We may ask, then, what does Eq. 1 tell us about how environmental variables affect the clutch size of a population?

From Eq. 1, clutch size should be larger in species that have larger primary male/female sex ratios, greater juvenile or adult mortality rates (smaller $\sum_{\alpha}^{\omega} \lambda_x$), later age of first breeding (smaller $\sum_{\alpha}^{\omega} \lambda_x$), or smaller number of broods reared (smaller $\sum_1^n P_i$). The lifetime reproductive success is $\sum_{\alpha}^{\omega} \lambda_x \sum_1^n P_i$, in terms of number of broods reared. Thus, as lifetime reproductive success increases, the mean clutch size decreases.

Clutch size and length of the breeding season

Longer breeding seasons allow breeding birds to lay more replacement clutches (following clutch failure) and to rear more broods (Ricklefs 1973, Yom-Tov 1987, Rowley and Russell 1991), increasing their probability of breeding successfully (greater $c_1 s_1$, $c_2 s_2$, ..., $c_n s_n$) and, therefore, on average, rearing a greater number of broods ($\sum_1^n P_i$) compared with species with shorter breeding seasons. Thus, according to Eq. 1, populations with a long breeding season should have a large $\sum_1^n P_i$ and a small clutch, and those with a short breeding season should have a small $\sum_1^n P_i$ and a large clutch (Murray 1991a). Because populations at higher latitudes typically have shorter breeding seasons and larger clutches than populations at lower latitudes, the theory accounts for latitudinal variation in clutch size, and because populations in the southern hemisphere have longer breeding seasons than populations at comparable latitudes in the northern hemisphere, the theory accounts for the smaller clutches in the southern hemisphere.

The widespread house wren provides a within species comparison. Its clutch size increases with increasing latitude in both the northern and southern hemispheres (Young 1994a). In Costa Rica it has a clutch size of 3.5 eggs and rears between zero and four broods during a breeding season of from seven to ten months (Skutch 1953, Young 1994b,

1996). Females on average rear two broods per year (Young 1994b). House wrens in Ohio, in the United States, rear on average one brood per year (range, 0 to 3) and have a mean clutch size of 5.3 eggs (Kennedy 1991) during a breeding season of 2.5- to 3-months (E. D. Kennedy, pers. comm.). Thus, the house wren in Costa Rica has a smaller clutch size and a longer breeding season, during which it rears more broods than does the house wren in Ohio.

The house sparrow *Passer domesticus* is another wide-ranging species, although much of its broad latitudinal range is a result of recent introductions (Summers-Smith 1988). Data are too few in the southern hemisphere for comparison. Summers-Smith (1988), however, showed that with increasing latitude within the northern hemisphere the breeding season shortens, the number of clutches laid per year decreases, and the clutch size increases. Nevertheless, the number of young reared per year per female does not change with latitude.

A more severe test of the theory is to explain the exceptions, for example, clutch-size differences among species at the same latitude. According to theory, clutch size should vary at the same latitude if the length of the breeding season varies among species at that latitude (Murray 1991a). I have already (Murray 1985, 1991a) noted that the mourning dove *Zenaida macroura* is a temperate North American species, although not a passerine, with a relatively small clutch size (2 eggs) and a relatively long breeding season (8-9 months) during which it rears an average of three broods (McClure 1942) and a maximum of six (Nice 1957). Most temperate species of the northern hemisphere have shorter breeding seasons during which an individual may rear no more than one or two broods. For example, prairie warblers *Dendroica discolor* in Indiana rear 0.77 (maximum = 2 [occasionally]) broods per female per year and have a mean clutch size of 3.89 eggs (Murray and Nolan 1989) during a 2.1-month breeding season (Nolan 1978). Florida scrub-jays *Aphelocoma coerulescens* rear 0.59 (maximum = 2 [very rarely]) broods per female and have a mean clutch size of 3.33 eggs (Murray *et al.* 1989) during a 3-month breeding season (Woolfenden and Fitzpatrick 1984). As already mentioned, house wrens in Ohio rear 1.08 (maximum = 3) broods per year and have a mean clutch size of 5.3 eggs (Kennedy 1991) during a 2.5- to 3-month breeding season (E. D. Kennedy, pers. comm.). The relationship between length of breeding season, number of broods, and clutch size is not perfect because in natural populations other factors also affect clutch size in each species (such as life expectancy and age of first breeding; see Eq. 1). For example, the house wren has a shorter life expectancy than the prairie warbler and Florida scrub-jay, and the latter begins breeding at a later age than the other two species. Nevertheless, the three passerines with the shorter breeding seasons rear fewer broods and have larger clutch sizes than the mourning dove, which has the longer breeding season.

The same relationship between clutch size, number of broods reared, and length of breeding season occurs within the tropics. The yellow-throated euphonia *Euphonia hirundinea* at Monteverde, Costa Rica, has a short breeding season for a tropical species, 50 of 52 clutches being laid in April, May, and June (Sargent 1993), compared with a mean length of breeding season of 6.6 months for birds in Costa Rica (Ricklefs 1966, 1969). The length of a breeding episode (i.e., about 39 days) plus post-fledging care of several weeks relative to that of the egg-laying season (i.e., less than 90 days) makes double brooding difficult (Sargent 1993). With a probability of nest success (s_1) of 0.40 (only 16 of 40 first-brood clutches reared any young), the probability of any female rearing two broods is small. Indeed, the species seems to be single-brooded (the only known nesting attempt following a success in rearing a brood failed; Sargent 1993).

Relatively large clutch sizes of several tropical passerine species also occur in southwestern Ecuador (Marchant 1960): e.g., short-tailed field-tyrant *Muscigralla brevicauda*, 3.96; snowy-throated kingbird *Tyrannus niveigularis*, 3.23; long-tailed mockingbird *Mimus longicaudatus*, 3.89; tropical gnatcatcher *Polioptila plumbea*, 3.23; crimson finch *Rhodospingus cruentus*, 3.21. In this region, the main breeding period varies from 6 to 14 weeks, being affected by seasonal and annual fluctuations in rainfall (Marchant 1959). Except for the mockingbird, second broods were unproved when suspected and, thus, were at best infrequent in these species (Marchant 1960).

Finally, in a less well-documented story, Lack and Moreau (1965) showed that passerines in tropical Africa had larger clutches in savanna habitat than in adjacent evergreen forest habitat. They hypothesized that breeding birds in the savanna had access to greater amounts of food relative to the size of the breeding population than did breeders in the forest, "since evergreen forest provides particularly uniform conditions throughout the year, whereas in savanna and other drier habitats, there is a flush of food with the restricted rainy season" (Lack 1968: 168). More uniform conditions may result in a longer breeding season than occurs in habitats subjected to alternating wet and dry seasons. I predict that the breeding season is longer in the more even climate of the evergreen forests than in the drier savanna habitats. More substantive data on the length of breeding seasons, number of broods reared, and clutch size from tropical habitats seem indicated.

A similar relationship between clutch size and habitat occurs in subtropical eastern Australia, where clutch size is greater in semiarid habitats than in wet habitats (Kikkawa 1974). Yom-Tov (1987) reported that species in the semiarid Eyrean zone had a greater mean clutch size (2.79 eggs) than species in the wet Irian and Torresian zones (2.18 eggs). The breeding season, however, is only slightly shorter (5.22

months) in the former than in the latter (5.48 months) (Yom-Tov 1987). Most of these, however, are comparisons of means of means of clutch sizes estimated from ranges. It would be desirable to have species-specific comparisons and have hard calculations of mean clutch sizes and egg-laying periods.

For example, Rowley and Russell (1991) provide hard data on three species of Australian birds (Table 1). The white-breasted robin *Eopsaltria georgiana* has the longest breeding season and the smallest clutch size, as expected from the theory. Among *Malurus* species, however, the splendid fairy-wren *M. splendens* has a longer breeding season and greater number of nests/female per year, fledglings per nest, and fledglings per year, but it has a larger clutch size than the red-winged fairy-wren *M. elegans*. According to the clutch size theory, the clutch size of *M. elegans* is smaller because it has much greater survival of fledglings to their first birthday and of breeding females, that is, it has a greater $\sum_{\alpha}^{\omega} \lambda_x$, about 2.9 versus 2.0. This comparison again shows that the relationship between length of breeding season, number of broods, and clutch size is not perfect because in natural populations other factors also affect clutch size in each species (Eq. 1).

What is interesting at this stage in the development of theory is that the predictions are consistent with all of the data at hand. We should, however, be increasing the quality of the soft data. Much of these data are available in the long-term studies undertaken during the past 30 years, but they are not being calculated or reported.

Table 1. Selected data on three species of Australian birds (from Rowley and Russell 1991).

	Malurus splendens	*Malurus elegans*	*Eopsaltria georgiana*
Breeding season	Sep.-Dec.	Oct.-Dec.	July-Dec.
Clutch size	2.9	2.4	1.9
Fledglings/nest	1.4	1.3	1.5
Nest/female per year	2.1	1.5	2.0
Fledglings/female per year	2.8	2.5	3.1
Survival (%)			
Fledgling to one year	34	41	28
Breeding female	66	80	73

Discussion

As Rowley and Russell (1991) pointed out, essential demographic data, especially the incidence and frequency of repeat clutches following either clutch failure or success (which I signify as c_1, c_2, ..., c_n) and the number of fledglings reared per female ($\sum_1^n P_i k_i$, where k_i is the mean number of young reared from a brood-i clutch), are not readily available in the literature, making evaluation of alternative evolutionary interpretations difficult. Future research should be directed toward the calculation and reporting of direct measurements of clutch size and c_i, s_i, and k_i for each i brood. With these we could then calculate $\sum_1^n c_i s_i$ $(=\sum_1^n P_i)$, which is the equivalent of annual reproductive success, counted as mean number of broods reared by the females of a population, and $\sum_{1}^{n} P_i k_i$, which is annual reproductive success, counted as the mean number of fledglings reared per female per year.

We should also want to have some indication of juvenile and adult survival and age of first breeding, affecting $\sum_{\alpha}^{\omega} \lambda_x$, because two populations could have breeding seasons of the same length but different clutch sizes as a consequence of different life expectancies or ages of first breeding (Eq. 1). For example, two tropical species with long breeding seasons, multiple broods, and large clutch sizes are the red-billed firefinch *Lagonosticta senegala* in Senegal (Morel 1964) and the large cactus finch *Geospiza conirostris* in the Galápagos (Grant and Grant 1989). Despite a breeding season of about nine months during which it rears as many as four broods (presumably large $\sum_{1}^{n} P_i$), the red-billed firefinch has a large clutch—four eggs (Morel 1964). This species, however, suffers great adult mortality, perhaps the greatest among birds—70 to > 75% per year (Morel 1964; Peach et al. 2001). In addition, there is large nestling mortality because of brood parasitism by the village indigobird *Vidua chalybeata*. Both result in a small $\sum_{\alpha}^{\omega} \lambda_x$.

In the large cactus finch (and other Galápagos finches), breeding success varies considerably with the amount of rainfall (Grant and Grant 1989). The breeding season varies from zero to eight months long, during which females may rear from zero to seven broods. In an average year, the breeding season is three to four months—shorter than average for a tropical species. In addition to occasional poor breeding seasons (i.e., no breeding), females have short life expectancies. Although a few females of the 1976 cohort survived to age 11, no females known to have fledged between 1978 and 1983 survived to age seven (Grant and Grant 1989). Only a few survived more than two breeding seasons. Furthermore, although females may breed as early as three months of age, females sometimes do not begin breeding until the end of their second year or later because they were born in a year preceding a poor breeding season (Grant and Grant 1989). Unfortunately, neither $\sum_1^n P_i$ nor $\sum_\alpha^\omega \lambda_x$ has been calculated for this species.

We should also want a more precise and accurate description of length of the breeding season. In the context of the problems discussed here, I suggest that investigators report the length of the egg-laying period for each species, a suggestion made earlier by Baker (1938). Calculating the number of equally good months for breeding by counting the proportion of clutches started in each month over a period of years for all species in a region (Thomas 1974, Wyndham 1986) could be misleading if the breeding schedules of several species overlap in time. Such a comparison could reflect average differences between regions but would not reflect the average length of breeding season of the species involved, much less the actual length of breeding season for any species.

Not everyone will be in a position to provide a complete demographic description of a population or species. Nevertheless, even a report on the number of fledglings reared per female should be useful. Virtually all long-term studies of populations of marked birds must have these data, but the authors fail to report them. The frequent practice of calculating the fraction of eggs that produce young that leave the nest (egg success) or the fraction of clutches that produce young that leave the nest (clutch success) or even the mean number of young that leave a successful nest may be misleading indicators of average reproductive success of the females involved (Murray 2000b).

My theory provides a consistent explanation for the increase in clutch size with increasing latitude, the difference in mean clutch size at comparable latitudes in the northern and southern hemispheres, the difference in clutch size among species at the same latitude, and much else (Murray 1979, 1985, 1991a, 1999, 2001b). What is needed are some high quality data for not only testing this theory but for exploring in greater detail the evolution of other features of life histories. Furthermore, it should be an interesting exercise to see if other theories on the evolution of clutch size could account for these clutch-size variations. So far, they have not (Martin et al. 2000).

Acknowledgements - I thank P. R. Grant, D. J. T. Hussell, J. Burger, S. Sargent, and especially J. R. Jehl, Jr., for commenting on an earlier version of the manuscript, and E. D. Kennedy for providing further information on the house wren.

Literature Cited

Baker, J. R. 1938. The relation between latitude and breeding seasons in birds. - Proc. Zool. Soc. Lond. 108A: 557-582.

Grant, B. R. and Grant, P. R. 1989. Evolutionary dynamics of a natural population: the large cactus finch of the Galápagos. - University of Chicago Press, Chicago, IL.

Hussell, D. J. T. 1985. Clutch size, daylength, and seasonality of resources: comments on Ashmole's hypothesis. - Auk 102: 632-634.

Johnson, A. W. 1965. The birds of Chile and adjacent regions of Argentina, Bolivia and Peru. Vol. I. - Platt Establecimientos Gráficos, Buenos Aires.

Johnson, A. W. 1967. The birds of Chile and adjacent regions of Argentina, Bolivia and Peru. Vol. II. - Platt Establecimientos Gráficos, Buenos Aires.

Kennedy, E. D. 1991. Predicting clutch size of the house wren with the Murray-Nolan equation. - Auk 108: 728-731.

Kikkawa, J. 1974. Comparison of avian communities between wet and semiarid habitats of eastern Australia. - Aust. Wildl. Res. 1: 107-116.

Klomp, H. 1970. The determination of clutch-size in birds. A review. - Ardea 58: 1-124.

Lack, D. 1947. The significance of clutch size. Parts I and II. - Ibis 89: 302-352.

Lack, D. 1948. The significance of clutch size. Part III. - Ibis 90: 25-45.

Lack, D. 1968. Ecological adaptations for breeding in birds. - Methuen, London.

Lack, D. and Moreau, R. E. 1965. Clutch-size in tropical passerine birds of forest and savanna. - L'Oiseaux 35 (special number): 76-89.

Marchant, S. 1959. The breeding season in s.w. Ecuador. - Ibis 101: 137-152.

Marchant, S. 1960. The breeding of some s.w. Ecuadorian birds. - Ibis 102: 349-382, 584-599.

Martin, T. E., Martin, P. R., Olson, C. R., Heidinger, B. J. and Fontaine, J. J. 2000. Parental care and clutch sizes in North and South American birds. - Science 287: 1482-1485.

McClure, H. E. 1942. Mourning dove production in southwestern Iowa. - Auk 59: 64-75.

Moreau, R. E. 1944. Clutch-size: a comparative study, with special reference to African birds. - Ibis 86: 286-347.

Morel, M.-Y. 1964. Natalité et mortalité dans une population naturelle d'un passereau tropical, le *Lagonosticta senegala*. - Terre et Vie 3: 436-451.

Murray, B. G., Jr. 1979. Population dynamics: alternative models. - Academic Press, New York.

Murray, B. G., Jr. 1985. Evolution of clutch size in tropical species of birds. - In: Buckley, P. A., Foster, M. S., Morton, E. S., Ridgely, R. S. and Buckley, F. G. (eds.), Neotropical ornithology, American Ornithologists' Union, Washington, D.C., pp. 505-519.

Murray, B. G., Jr. 1991a. Sir Isaac Newton and the evolution of clutch size in birds: a defense of the hypothetico-deductive method in ecology and evolutionary biology. - In: Casti, J. L. and Karlqvist, A. (eds.), Beyond belief: randomness, prediction, and explanation in science, CRC Press, Boca Raton, Florida, pp. 143-180.

Murray, B. G., Jr. 1991b. Measuring annual reproductive success, with comments on the evolution of reproductive behavior. - Auk 108: 942-952.

Murray, B. G., Jr. 1992. The evolution of clutch size: a response to Wootton, Young, and Winkler. - Evolution 46: 1581-1584.

Murray, B. G., Jr. 1999. Predicting the occurrence of synchronous and asynchronous hatching in birds. - In: Adams, N. J. and Slotow, R. H. (eds.), Proc. 22 International Ornithological Congress, Durban, South Africa, pp. 624-637. (Published on CD)

Murray, B. G., Jr. 2000a. Universal laws and predictive theory in ecology and evolution. – Oikos 89: 403-408.

Murray, B. G., Jr. 2000b. Measuring annual reproductive success in birds. - Condor 102: 470-473.

Murray, B. G., Jr. 2001a. Are ecological and evolutionary theories scientific? - Biological Reviews 76: 255-289.

Murray, B. G., Jr. 2001b. The evolution of passerine life histories on oceanic islands, and its implications for the dynamics of population decline and recovery. – Studies in Avian Biol. 22: 281-290.

Murray, B.G., Jr., and Nolan, V., Jr. 1989. The evolution of clutch size. I. An equation for predicting clutch size. - Evolution 43: 1699-1705.

Murray, B.G., Jr., Fitzpatrick, J. W. and Woolfenden, G. E. 1989. The evolution of clutch size. II. A test of the Murray-Nolan equation. - Evolution 43: 1706-1711.

Nice, M.M. 1957. Nesting success in altricial birds. - Auk 74: 305-321.

Nolan, V., Jr. 1978. The ecology and behavior of the prairie warbler *Dendroica discolor*. – Ornithol. Monogr. 26. Allen Press, Lawrence, KS.

Peach, W. J., Hanmer, D. B. and Oatley, T. B. 2001. Do southern African songbirds live longer than their European counterparts? – Oikos 93: 235-249.

Ricklefs, R.E. 1966. The temporal component of diversity among species of birds. - Evolution 20: 235-242.

Ricklefs, R.E. 1969. The nesting cycle of songbirds in tropical and temperate regions. - Living Bird 8: 165-175.

Ricklefs, R.E. 1973. Fecundity, mortality, and avian demography. - In: Farner, D.S. (ed.), Breeding biology of birds, National Academy of Sciences, Washington, D.C., pp. 366-435.

Ricklefs, R.E. 1980. Geographical variation in clutch size among passerine birds. - Auk 97: 38-49.

Rowley, I. and Russell, E. 1991. Demography of passerines in the temperate southern hemisphere. - In: Perrins, C. M., Lebreton, J. D. and Hirons, G. J. M. (eds.), Bird population studies: relevance to conservation and management, Oxford University Press, Oxford, pp. 22-44.

Sargent, S. 1993. Nesting biology of the yellow-throated euphonia: large clutch size in a neotropical frugivore. - Wilson Bull. 105: 285-300.

Skutch, A.F. 1949. Do tropical birds rear as many young as they can nourish? - Ibis 91: 430-455.

Skutch, A.F. 1953. Life history of the southern house wren. - Condor 55: 121-149.

Skutch, A.F. 1985. Clutch size, nesting success, and predation on nests of neotropical birds, reviewed. - In: Buckley, P. A., Foster, M. S., Morton, E. S., Ridgely, R. S. and Buckley, F. G. (eds.), Neotropical ornithology, American Ornithologists' Union, Washington, D.C., pp. 575-594.

Summers-Smith, J.D. 1988. The sparrows: a study of the genus *Passer*. - T&AD Poyser, Calton, England.

Thomas, D.G. 1974. Some problems associated with the avifauna. In: Williams, W.D. (ed.), Biogeography and ecology of Tasmania, Junk, The Hague, pp. 339-365.

Woolfenden, G. E. and Fitzpatrick, J. W. 1984. The Florida scrub jay: demography of a cooperative-breeding bird. - Princeton Univ. Press, Princeton, New Jersey.

Wootton, J. T., Young, B. E. and Winkler, D. W. 1991. Ecological versus evolutionary hypotheses: demographic stasis and the Murray-Nolan clutch size equation. - Evolution 45: 1947-1950.

Wyndham, E. 1986. Length of birds' breeding seasons. - Am. Nat. 128: 155-164.

Yom-Tov, Y. 1987. The reproductive rates of Australian passerines. - Austral. Wild. Res. 14: 319-330.

Yom-Tov, Y., Christie, M. I. and Iglesias, G. J. 1994. Clutch size in passerines in southern South America. - Condor 96: 170-177.

Young, B. E. 1994a. Geographic and seasonal patterns of clutch-size variation in house wrens. - Auk 111: 545-555.

Young, B. E. 1994b. The effects of food, nest predation and weather on the timing of breeding in tropical house wrens. - Condor 96: 341-353.

Young, B. E. 1996. An experimental analysis of small clutch size in tropical house wrens. - Ecology 77: 472-488.

Commentary (*Oikos*, 18 February 2002)

This paper has a long history. In reviews received from 8 journals, only one reviewer was enthusiastic about having my paper published. Lengthy commentaries by me on the reviews were to no avail. The diversity of criticism is extraordinary, but, although I think anyone interested in clutch-size evolution should read them, I am not going to reprint them here (takes up too much space). I have reproduced here my commentary on the *Oikos* (the last) version.]

Dear …

Here we go again.

Apparently, the most important point made by the reviewers, since it made it into your letter of rejection, is (quoting from your letter), "much of what is said here is already published (in Biol. Rev. 76)." That is a real stretch of the truth. In the *Biological Reviews* paper, I wrote the following with regard to clutch size and the length of the breeding season:

Now, suppose that $\sum_1^n P_i$ is directly correlated with the length of the breeding season because where breeding seasons are long females are able to produce more replacement clutches following clutch failure and are able to rear more than one brood per year than females living where breeding seasons are short. The prediction is that, because $\sum_1^n P_i$ is greater where breeding seasons are longer, the clutch size should be smaller where breeding seasons are longer. This is a relationship unknown to ornithologists but predicted by me (Murray, 1979, 1985*a*, 1991*b*). The prediction is confirmed by clutch size being smaller at tropical latitudes than at higher latitudes (Lack, 1947, 1948*a*, 1954, 1968; Cody, 1966, 1971; Klomp, 1970; Skutch, 1985) and smaller in the southern hemisphere than in the northern (Moreau, 1944; Yom-Tov, 1987; Rowley & Russell, 1991; Yom-Tov, Christie & Iglesias, 1994) because breeding seasons are longer in the tropics than at higher latitudes and at southern latitudes than at comparable northern latitudes (Baker, 1938; Skutch, 1950; Wyndham, 1986; Rowley & Russell, 1991; and Yom-Tov *et al.*, 1994). It is confirmed further by clutches being larger in species with short breeding seasons within the tropics (Marchant, 1960; Sargent, 1993). I have reviewed the evidence in greater detail elsewhere (Murray, in preparation). Here, we have a prediction of an empirical relationship that was unrecognized by ornithologists before the prediction was made, and the prediction (so far) is consistent with empirical evidence.

Try as I might I can find no discussion of any case in the preceding paragraph. What I did at that point in my *Biological Reviews* paper was pontificate (after all, it was a review of which the evolution of clutch size was a small part). I made an argument without presenting the evidence (other than a list of citations—and, don't forget, ecologists evidently do not read the cited literature). My evidence is in the paper "(Murray, in preparation)," which is how the editor wanted me to refer to the paper that I had submitted elsewhere and now submit to you. In MS #11623, besides showing how my theory accounts for the well-known, if not universal, increase in clutch size with increasing latitude and the smaller clutch sizes in southern hemisphere species compared with northern hemisphere species at comparable latitudes, which has never been explained, I discussed the evidence and arguments produced by others on the topic, the geographic variation in the few species with wide distribution (House Wren, House Sparrow); *large* clutches in the tropics (*Euphonia, Lagonosticta, Geospiza,* and other species in SW Ecuador), which have never even been noticed as important exceptions, much less explained, by anyone; clutch-size variation in Australia; and some specific, detailed comparisons of related species in Australia.

It is always a cheap shot to say that a submitted ms. simply repeats what has already been published, but it is much harder for anyone to make that charge stick when it is untrue. The reviewer who suggested that much of my ms. (4,050 words) has already been published in my *Biological Reviews* paper (236 words) has written an untruth.

Now to the specific points made by the reviewers:

The shorter review (the one on the smaller piece of paper)

(1) The reviewer writes, "the asssociation [sic] between the length of the breeding season and clutch size was originally proposed by Ricklefs." Did I not cite Ricklefs? Let's see now, what did I write about Ricklefs? Here it is, "Longer breeding seasons allow breeding birds to lay more replacement clutches (following clutch failure) and to rear more broods (Ricklefs 1973, Yom-Tov 1987, Rowley and Russell 1991)." If you read Ricklefs, as I have, you will find only that, despite the fact that "few data are available on the number of clutches laid, or broods raised, by multiple-brooded species," he nevertheless discusses the effect of the length of the breeding season on the number of clutches started and broods reared. He makes no reference to clutch size, and he sees no theoretical importance to the facts that he reports.

(2) The reviewer seems compassionate, "The author's aim is a legitimate one, BUT [my emphasis] the justification for publication of a paper is in some new information or hypothesis" (Say what?—Not a new synthesis of data in

support of a predictive theory? No new explanations for previously unexplained phenomena? These are not justifications?). I suggest to you that the justification of my paper is the simple fact that I discuss in detail ideas that the great majority (even if we count only the ornithologists) of your readers are completely unaware. As I noted in my ms., "Martin et al. (2000) pointed out (italics added by me here), 'The inability of the most widely invoked hypotheses to explain latitudinal patterns in clutch size illustrate [sic] that alternative hypotheses … *deserve more attention, and that current theories of clutch size evolution need major revision.*' Martin et al. (2000), however, did not consider my theory on the evolution of clutch size (Murray 1979, 1985, 1991a), which predicts the clutch-size variations just described." Just maybe, ornithologists should give my hypothesis a look. After all, no one has, during the past 23 years, found a flaw in my argument or offered a counterexample.

(3) Your reviewer writes, "The few examples provided in the present paper hardly justify the publication of the present manuscript." This statement emphasizes the differences in philosophy between me and your reviewer, as I described in the *Biological Reviews* paper. Your reviewer is a Baconian, inductive verifier. He wants many more examples, as if all the reported patterns in geographic variation of clutch sizes of hundreds, if not thousands, of species are not enough. I am a Popperian, deductive, unifier. A critic of me needs to find (in naive falsification) *one* counterexample. He cannot find one, so he (and presumably the subeditor and editor) demands that I explicitly identify all of the many species that have been considered by Lack, Klomp, and Cody in their reviews of clutch-size variation with latitude, a pattern that is unchallenged by anyone. I think this is just being obstructive. It is this kind of thinking that convinces me that ecology and evolutionary biology are not sciences.

(4) The reviewer states, "The author says that he eschew [sic] correlations, but at present his hypothesis is not proven, and in order to support it he presents correlations." Again the reviewer betrays his scientific philosophy, which differs considerably from mine (see *Biological Reviews* paper again). In real science, hypotheses are *never* proved. Furthermore, I do not understand what he is writing about correlations. I do not rely on statistical correlations to support my case—that is, I do not argue, "X is statistically different from Y, therefore my theory must be true." Indeed, my theory predicts and, therefore, explains the correlations and patterns discovered by others—there are no exceptions that are masked in a "correlation." My theory explains the exceptions to the correlations (e.g., large clutches in the tropics).

(5) The reviewer states, "Please explain what is "Malthusian parameter." Is he kidding me? Does he not know what the Malthusian parameter is, or does he want me to define it again? (I think it is the former.)

(6) The reviewer states, "The author states categorically his 3 laws, without explaining them here. I question the validity of the third law." Well, just how many times do I have to repeat the meanings of my three laws for those putative "experts" (my so-called "peers") who choose not to read the cited literature? So, the reviewer "questions" the validity of my third law. I can live with that, but what is his question? If he thinks (oops! I mean believes) that my third law is unjustifiable, why does he believe that? What I find difficult to live with is those ecologists who think I am wrong but cannot articulate a counterargument or even a question or comment other than, "I don't believe it, therefore it must be untrue." The reviewer is in this category. Why should anyone have any confidence in an anonymous person's unsupported musings?

(7) The reviewer states, "Please explain why Equation 1 'must be true,'" Good grief! Does this reviewer not want to

read any of the cited literature? If I have to justify every sentence in this paper that I, and others, have justified elsewhere, the paper will be much longer than what I have already written (and, of course, I will be writing what I and others have published elsewhere).

In this paper all you have to know is that I made the prediction earlier. If a reader wants to check the prediction, he should read the cited literature. The questions regarding this paper should be, "Are the examples accurately presented?" and "Are there any counterexamples that the author has ignored?" For me to have to write again all the background information needed for an ignorant (regarding the topic under discussion) reader may need to understand what is going on really does seem unjustified. An author cannot write his paper for the most uninformed person who might choose to read it.

(8) The reviewer states, "Most Columbiformes lay a fixed clutch of 2 eggs (some only 1), and comparing a dove to passerines is a mistake." This reviewer, like other ecologists, is enamored with pontification—making statements without supplying evidence or argument (or counterargument to my explicitly stated argument). He says that comparing doves with passerines is a mistake, so we are supposed to be impressed and accept what this anonymous, apparently uninformed reviewer has to say. My theory, as I have stated several times, but unknown to the reviewer, is applicable to all organisms, not just passerines. The laws of population dynamics are applicable to all species. They are not different for different species. Where is the substance in this review? The reviewer is uninformed with regard to my theory, the facts, the philosophy of science, and much else.

The longer review (the one on the larger piece of paper)

(1) This reviewer starts off with, "This manuscript addresses the question: why does the clutch size increase

with increasing latitude?" No, it does not. It addresses the inverse relationship between clutch size and the length of the breeding season. Not only does it explain the clutch size increase with increasing latitude, but it explains why clutch sizes in the southern hemisphere are smaller than at comparable latitudes in the northern hemisphere, not previously explained. It explains why the clutch size is larger in some tropical species than those of other species at the same latitude, not previously explained. The theory itself explains much else. This reviewer does not make a good start.

(2) The reviewer continues (I added the italics), "The author points out that annual reproductive effort can roughly [?, BGM] be taken as [average clutch size]*[average number of clutches per year (reviewer's brackets here)], which means that if a bird increases the number of clutches/year, it must decrease the clutch size *if it is to keep the annual reproductive effort constant.*" This is not what I have written, nor is it what I think. What I say is, if the average female of a population is able to rear successfully more than one brood during a breeding season, then we should expect smaller clutch sizes in that population because "Selection favors those females that lay as few eggs or bear as few young as are consistent with replacement because they have the highest probability of surviving to breed again, their young have the highest probability of surviving to breed, or both." Nowhere do I suggest that there is any need on the part of females to keep a constant annual reproductive effort. It is this kind of misreading, subtle as it may seem, that prevents readers from understanding what I have written.

(3) The reviewer writes, "To find out whether [birds with long breeding seasons have lower clutch sizes] is so or not, is an empirical problem. The problem with the manuscript is that empirical data to test this idea is [sic] missing." Say what?? Has anyone (reviewers, subeditors, editors) read this paper? First, I point out that one well-

known and unquestioned (as far as I know) correlation between clutch size and latitude (increasing clutch size with increasing latitude) is also an inverse correlation between clutch size and length of breeding season (a proposition not previously recognized by ornithologists). Second, I pointed out that the clutch sizes in the southern hemisphere, which are smaller than at comparable northern latitudes (a well-known fact that has been unexplained) are also examples of the inverse correlation between clutch size and length of breeding season because the breeding seasons of birds are longer in the southern hemisphere than at comparable latitudes in the northern hemisphere. I point out that the clutch size of *Euphonia* in Costa Rica and several species studied by Marchant in SW Ecuador are large despite being tropical because their breeding seasons are short compared with the long breeding seasons typical of tropical species. I discussed within-species variation in the House Wren and the House Sparrow, in both of which clutch size is inversely correlated with latitude. I ask again, Has anyone read my paper?

(4) In this passage, the reviewer surmises, "A few isolated cases (house wren, and references to a published paper on african [sic] passerines, and work on australian [sic] birds, where the relation does not seem to hold anyway) is all we get." I really did not want to get into it in detail because Yom-Tov (1987) used means of mean clutch sizes and means of mean length of breeding seasons for comparisons, a seriously misleading practice. If one looks at an avifauna as a whole, there may be breeding year round, when in fact each species may breed for no more than three or four months. But your reviewer accepts Yom-Tov's ambiguous data and Yom-Tov's conclusion that the relationship between clutch size and latitude does not hold in Australia. Yom-Tov, of course, is acceptable because he does not challenge the conventional wisdom. The data presented by Yom-Tov do not preclude the relationships I describe.

(5) The reviewer writes, "To find out if the pattern of clutch size vs. latitude has anything to do with breeding seasons, one has to compare the breeding season at different latitudes across several hundreds of species." I wonder what the reviewer thinks that Baker, Lack, Marchant, Klomp, Cody, Skutch, Kikkawa, Rowley and Russell, Yom-Tov (Australia), Yom-Tov et al. (South America), Wyndham, and others have done. The reviewer stated, "I would not be so sure however that data on this would be impossible to find." It can be found in the literature cited in my paper, if the reviewer would read it.

(6) The reviewer states, "There is a fair bit of unnecessary mathematics in the manuscript, and it does little to either support or refute the idea." The mathematics was not intended to support or refute the idea. It was intended to give readers the necessary quantitative relationships (apparently unknown to ornithologists) to think about clutch size.

Furthermore, the mathematics in the paper consists of a single equation. By "unnecessary mathematics," the reviewer must be referring to my use of $\sum_{\alpha}^{\omega} \lambda_x$ to refer to "mean number of breeding seasons for females from successful clutches (that is those that produce at least one young to leave the nest)" and $\sum_{1}^{n} P_i$ to refer to "the number of broods reared per female during a breeding season." If I had not used those symbols I would have had to write out the phrases each time you see those symbols in the ms. Is it too much to ask readers to remember the definitions of two symbols?

(7) The reviewer states, "[The mathematics] cannot [either support or refute the idea], for equation (1) gives only the expected clutch size in a constant population." This is untrue. I suggest that the equation is applicable to any naturally occurring population, and no one has offered an example indicating why it is not, despite the fact that they

have the entire plant and animal kingdoms in which to search for an exception. He continued, "There are a number of unstated assumptions that has to hold for (1) to be true, no age-related variation in either clutch size or annual clutch number, and no differences in state variables." This is untrue—the word "mean" cannot be inferred to mean "constant" (ecologists seem to be stumped by this concept—see my letter to you of 14 February). The clutch-size equation predicted pretty well the clutch sizes of the Prairie Warbler, Florida Scrub Jay, and House Wren, the *only* species that have been examined in this way, all of which have a variable clutch size. Had other ornithologists tried to calculate their population's mean clutch size from the data they had available, we would have a better understanding of both clutch size and the clutch-size equation. I wrote to about 25 investigators, requesting that they undertake such an analysis and did not even receive an acknowledgment. I am distressed at how easily ecologists can believe and, worse, write untruths without any evidence or argument. They can do this with equanimity because they are not going to be held accountable.

(8) The reviewer writes, "the *Malurus* species referred to are cooperative breeders, and this aspect of their breeding behaviour is very likely to influence their clutch size." This looks like more pontification. The reviewer *believes* that cooperative breeding should influence the clutch size and implies that the clutch-size equation would be inadequate because it assumes "no differences in state variables," and "Differences in state are always relevant in life-history optimality." What the reviewer does not mention is that the Florida Scrub Jay is a cooperative breeder and, nevertheless, provides the best example showing the rigorousness of the clutch-size equation (predicted clutch size = 3.43, actual clutch size = 3.33; Murray et al. [1989]). This reviewer seems not to know very much about clutch size in birds.

(9) The reviewer continues, "given the assumptions are true eq(1) must support any theory for the pattern." What the reviewer does not mention is that Murray and Nolan (1989) wrote, "We believe that [the clutch-size equation] expresses the consequences of selection favoring females that lay as few eggs as possible in a clutch while replacing themselves, and we consider the fact that the equation was derived from the hypotheses of the theory as support for the theory. Nevertheless, we recognize the possibility that [the clutch-size equation] is no more than a description of a dynamic relationship that has evolved in some other way. But that other way remains to be described and explained." Murray (1991) wrote, "[The clutch size equation] must hold, whether one accepts Lack's hypothesis, Skutch's hypothesis, my hypothesis, or some other hypothesis as the best explanation for the evolution of clutch size." Murray (1992) wrote, "[The clutch-size equation] is a description of the relationship between clutch size and other demographic parameters, which should be true for any theory on the evolution of clutch size, whether one assumes the clutch size to be maximized, optimized, or minimized by natural selection. ... the relationship of the equation to other hypotheses on clutch-size evolution has not been described and explained (Murray and Nolan 1989)."

I have never claimed that the clutch-size equation itself supported my theory. The equation merely provides us a means of calculating clutch size from other demographic data and a means of thinking rigorously about the evolution of clutch size. For example, it seems pretty clear that the conventional wisdom regarding the role of predation on clutch size—the greater the predation, the smaller the clutch size—is mathematically impossible. Nevertheless, Tom Martin (and I presume his reviewers and editors) continues to promote this proposition without any reference to my demonstration that it is false.

(10) The reviewer adds, "E.g., Ashmoles [sic] hypothesis is supported because survival appears in the denominator of the expression." No, it does not. The reviewer does not understand. If the clutch-size equation must be true regardless of anyone's clutch-size hypothesis, then it does not "support" any hypothesis (including mine). Inasmuch as Ashmole's hypothesis concerns the role of available food supply during the breeding season, I really do not see the relevance of survival appearing in the denominator. But that's another story, and another example of the reviewer's pontification—a statement made without either evidence or argument. I have pointed out, however, that the only theory I know of that is consistent with the clutch-size equation is mine. I should think that the promoters of other theories should be trying to show how the clutch-size equation is consistent with theirs.

(11) The reviewer states, "Finally, there is a theoretical reason why we might not expect breeding seasons to be important for the evolution of clutch size.

$\dfrac{\partial}{\partial n}\left(\sum_{1}^{n} P_i\right)$ measures the sensitivity of clutch size to a change in the number of annual clutches. Increasing n will give a decreasing response in clutch size, because subsequent Pi's [sic] are smaller. $\dfrac{\partial}{\partial \alpha}\left(\sum_{\alpha}^{\omega} \lambda_i\right)$ measures the sensitivity to the timing of onset of breeding season [?—I suppose he means age of first breeding]. A decreased alpha [age of first breeding] will have a strong influence on clutch size because the earlier lambdas in the sum are larger. ..."

The reviewer is pontificating here. He seems to be a mathematician rather than a biologist. Therefore, he misunderstands what the clutch-size equation is all about. He seems to interpret $\sum_{1}^{n} P_i$ and $\sum_{\alpha}^{\omega} \lambda_x$ to be mathematical functions of some sort, when in fact they are numbers obtained from data collected in the field. $\sum_{1}^{n} P_i$ is a number = (number of broods reared by a population of females)/(number of females of breeding age). $\sum_{\alpha}^{\omega} \lambda_x$ is a number that is the sum of values of λ_x taken from a life table. These are empirically determined numbers. There is no reason to calculate partial derivatives, even if you could, because the equation states that the empirical clutch size is calculated from those specific empirical numbers. The actual mean clutch size is that calculated with the equation, or it is not. The way to test whether the equation works is to calculate the clutch size from other empirically-determined data. The only empirical data available show that the equation works (Murray and Nolan 1989; Murray et al. 1989; Kennedy 1991). There are no examples showing that the equation does not work.

(12) The reviewer continues, "A decreased alpha will have a strong influence on clutch size because the earlier lambdas in the sum are larger. Whenever there is variation in both n and alpha, clutch size will be more influenced by the variation in alpha, which may easily overshadow any influence of n. This is of course again given that population is constant, and that the other necessary conditions hold."

The reviewer is lecturing (more pontification) to me (or you). In fact, I have discussed the effects of differences in age of first breeding and number of broods on clutch size several times (e.g., Murray 1979, 1984, 1991). Also, the population does not need to be "constant" (whatever that means—in size? in age structure?). All that is needed is a decent sample size. Ecologists are fearful of testing my theory because, if I am correct, they are going to have to admit that they believed something that was not true. So, they choose to throw up a smokescreen.

The reviewer seems not to understand the nature of either my theory or this paper. My theory is deductive-nomological, which means that one predicts clutch size and tests it by searching for the predicted empirical evidence. I have earlier made the prediction. In this paper I present the empirical evidence.

Unhappily, my experience is, if an editor requests that I respond to the reviewers' negative comments *before* he makes a decision, my papers always get published. If I respond to the reviewers' comments *after* an editor has made his decision, the paper almost never gets published. [Nevertheless, I am still hopeful that Professor Malmer will change his mind and allow me to respond to Turchin's egregious insulting treatment of my ideas. Turchin does not only disrespect me, he disrespects his readers, when he misrepresents the work of others.] This leaves me with a slight problem with regard to this paper, which contains only facts and no theory. It is a *fact* that I predicted from my laws and the clutch-size equation that there should be an inverse relationship between clutch size and length of the breeding season. It is a *fact* that this relationship was unknown to ornithologists (and apparently still is) prior to my prediction. It is a *fact* that the relationship exists, as revealed by the examples I have given in this paper. It is a *fact* that these examples have never been discussed in the literature with respect to my theory. It is a *fact* that no counterexamples have been given for *any* aspect of my theory. It is a *fact* that the reviewers, as a group, cannot read and cannot do the arithmetic. I think that I am on solid ground when I think that the reviewers and editors are interested in preserving the conventional wisdom of a faith-based belief system called ecology. I have not the slightest doubt that my papers do not get accepted because I am critical of the faith.

The question is, What can I do about it? The answer is, nothing. I have written a lot on population dynamics and the evolution of life histories, but I have not yet generated a dialogue with any so-called scientist colleague. There is an extensive correspondence on this paper. What am I to do? The only thing I can think of to do is to put the manuscript and the (20±) reviews generated by it on a Web page with my responses to all the criticism. This should certainly reflect the state of the field and the incompetence, as scientists, of those who have commented on my work. I know that I will not get any points from my colleagues for exposing ecology and evolution as they are practiced, but I am not getting any points from them by writing thoughtful and what should be stimulating papers, that is, stimulating to anyone without a closed mind.

If you can give me an alternative, I should really like to know of it, so that I could consider that option.

My greatest disappointment is that I have been submitting papers to *Oikos* for 18 years and reviewing manuscripts for the editors for almost as long without having gained either credibility or respect from the editors of *Oikos*. I thank the editors for publishing as many of my papers as they have (I am truly grateful), but maybe gaining credibility and respect would have meant more to me.

Cheers, Bert

Chapter 6

The Calculation of Clutch Size: The Mayfield method is Mathematically Flawed

[I have long wondered why ornithologists, ecologists, and evolutionary biologists have ignored my published work. My lament was the same as all those in the past who claimed that their work was "misunderstood"—that is, I believed that my colleagues could not "understand" what I was writing about, despite my efforts to provide a solid philosophical, mathematical, and empirical foundation for what I think. What has become apparent to me in the past decade or so is that my colleagues cannot understand high school algebra, despite all the mathematics and, especially, statistics that decorate their papers. I offer this chapter as evidence that my colleagues are innumerate. By the standards of Paulos [1988, *Innumeracy*], however, I too am innumerate, but I believe that I am able to do elementary mathematics—addition, subtraction, multiplication, division, and logarithms.

An important parameter in reproductive biology is clutch success, which, more often than not is referred to as "nest success." My preference is for clutch success because I am interested in the success of clutches rather than in the success of nests. The term "nest success" infers that some nests that were built but not laid in may be included in the analysis. Nevertheless, I use the terms, as does everyone else, interchangeably.

The calculation of clutch success seems straightforward. If we find 50 nests with clutches and 35 of them are successful, then the probability that a clutch will survive an entire nest episode from the laying of the first egg to the departure of the first fledgling (i.e., clutch success) is 35/50 = 0.70. Mayfield (1960, 1961) pointed out that in most studies most or many nests are found sometime after the date of first egg-laying. Mayfield argued that because later-found nests should have a higher probability of success than earlier-found nests, samples that included late-found nests should overestimate the overall probability of nest success. Mayfield's criticism of what has been called the "simple method," was completely justified. Mayfield (1960, 1961, 1975) proposed a method for calculating nest success from samples including clutches found after first egg-laying. The Mayfield method has been influential, there being over 1,000 published papers citing his three publications. Unfortunately, his solution was mathematically flawed.

I first met Harold Mayfield in 1961, when I started as a graduate student at the University of Michigan. We knew each other for years and spent about two weeks together in Oaxaca in 1979 with Bill Hardy studying finches. I can remember discussing the paper, and I can remember him mentioning that he had consulted a statistician. I no doubt read his paper, but I was not yet interested in calculating clutch success. My interest in clutch success was theoretical rather than empirical, but it became clear that, if I were to test my clutch-size theory, I would eventually have to refer to data. I finally read the Mayfield papers carefully. I was shocked. There are several rather obvious errors in high school algebra. I wrote a paper, which I thought the editor would send immediately to the printer. Imagine my surprise when the paper was rejected, not by one journal but at least perhaps times by six journals (I have forgotten the count.)! I was shocked! I had wondered why no one else had discovered these errors during the preceding 45+ years. I found out. Ornithologists are innumerate. This is demonstrated in their reviews.

My paper underwent several revisions and was submitted under different titles to different journals in my desperate search to find the combination of words and illustrative examples that could get through to innumerate reviewers and editors. My criticism of the Mayfield method, however, is the same in all of them—only the numerical examples differ. The manuscript reproduced here, "On calculating nest success: A critique of the Mayfield method," was sent to the *Condor* on 6 October 2008. It was rejected on 4 December 2008. As a result of the reviews, I have corrected this ms. These corrections have been explained in the ms. and a letter to the editor (Commentary 6.A.)]

MANUSCRIPT

On Calculating Nest Success: A Critique of the Mayfield Method

Abstract. An important measure of reproductive success is the probability that a nest will survive from the time eggs are first laid until the chicks leave the nest as fledglings (i.e., nest or clutch success). The simple method for calculating nest success is to divide the number of successful nests by the number of nests with known outcomes (either success or failure). Mayfield (1960, 1961, 1975) considered the simple method unsatisfactory and proposed a model for calculating nest success more accurately. In this paper I argue that the Mayfield method and other methods based on the counting of exposure days suffer from several mathematical errors, which I discuss in some detail: (i) the exposure days prior to nest discovery are not counted; (ii) the number of failures divided by exposure days is not mathematically equivalent to daily mortality rate; and (iii) the daily mortality or survival rate cannot be constant under any circumstances for any period of time (except for 100 percent survival per day). I conclude that the simple method applied to nests found at or before first-egg-laying provides the most accurate estimate of the nest success of a sample from a population.

Keywords: clutch success, daily mortality rate, daily survival rate, Mayfield method, nest success.

INTRODUCTION

The probability that a clutch will survive an entire nest episode from the laying of the first egg to the departure of the first fledgling (i.e., clutch success) is an important parameter of reproductive biology. The calculation of this probability seems straightforward. For example, with what has been called the "simple" method, if we have found 40 nests and 15 were successful, then the probability of success is $15/40 = 0.375$. Alternatively, if we have been successful in following each of the 40 nests daily from the day the first egg was laid until nest failure or success (at least one fledgling leaves the nest), then we could calculate the daily survival rate of nests (d_t), that is, the number of nests that survive from day t to day $t + 1$ divided by the number of nests at risk on day t (i.e., $d_t = N_{t+1}/N_t$). With these data, we could determine the probability of clutch success (s) with,

$$s = \prod_{0}^{f-1} d_t ,\qquad\qquad (1)$$

where f is the first day on which at least one young leaves the nest, and the first egg is laid on day 0. With this equation, d_t may vary from day to day. These relationships may better be understood by way of example. From an imaginary set of nest-record cards, let us enter on a spreadsheet (Table1) the day the nest was discovered (the first X in each row in Table 1), the subsequent days when eggs or young were found (also marked X), and the day of success (marked S) or failure (marked F). (Let us truncate the nest episode to 13 days [12 24-hour intervals] from laying of first egg to first fledging in order to save space in Table 1)

Given the data in Table 1, we count how many clutches have survived from day 0 (i.e., 40) to day 1 (i.e., 39)—the daily survival rate, d_0, is 39/40 = 0.9750. We next count the number that survived to day 2 (i.e., 36)—the daily survival rate, d_1, is 36/39 = 0.9231. We continue through the nest period of 13 days (Table 1). Using Eq. 1, s is 0.375.

This method requires the investigator to follow the sample nests throughout the nest episode from the laying of the first egg through the day of failure or to the day of first fledging. A problem arises in estimating clutch success when the investigator is unable to monitor the population every day.

In most studies nests are often first found after the first eggs are laid. Mayfield (1960, 1961) correctly argued that because nests found later in the nest episode should have a higher probability of success than nests found earlier, samples that included late-found nests should overestimate the overall probability of nest success. Mayfield (1960, 1961, 1975) proposed a method for calculating nest success from samples that include nests found later in the nest episode, using the concept of "exposure days." The Mayfield method stimulated the proposal of at least 21 other models for estimating nest success, which "offer a bewildering choice to a biologist posing a rather simple but important question— what is the success rate of a group of nests?" (Johnson 2007a: 67).

In this paper, I point out that the Mayfield method (most recently described succinctly by Johnson 2007b) and other methods using the concept of exposure days are fundamentally flawed and calculations made with them give incorrect values for daily survival rates for nests and for nest success. I discuss these errors in detail and recommend that the Mayfield method and related methods no longer be used.

Given the diversity of models and applications, I emphasize that I am concerned in this paper *only* with what Johnson (2007a) called the most important question—what is the probability that a nest in a sample of nests will survive a full nest episode from the laying of the first egg to the departure of the first fledgling from the nest.

THE MAYFIELD METHOD

Mayfield (1960, 1961, 1975) argued that the usual method at that time for calculating nest success (i.e., the simple method) provided a biased estimate because later-found-nests have a greater probability of success than earlier-found-nests, and proposed a method for eliminating the bias. He suggested that investigators should calculate the daily probability of nest success by first counting "exposure days" for each nest under observation. For example, a nest observed for five days between discovery and failure or success counts as five exposure days, and a nest observed for ten days between discovery and failure or success counts as ten exposure days. Mayfield's "daily mortality rate" (DMR) is the number of failed clutches divided by the total number of exposure days for the nests in the sample. DMR is subtracted from one to give the "daily survival rate" (DSR). Assuming that the DSR is a constant, the probability of nest success (s), according to the Mayfield method, is,

$$s = DSR^{f-1}, \tag{2}$$

where f is the day the first fledgling leaves the nest.

TABLE 1. Simulated spreadsheet of a sample of 40 nests, each of which was monitored from day of first laying to day of failure (F) or success (S).

Nest	Day of Nest Episode (x)[a]													Day[b]
	0	1	2	3	4	5	6	7	8	9	10	11	12	
1	X	X	X	X	X	X	X	X	X	X	X	X	S	12
2	X	X	X	X	X	X	X	X	X	F				9
3	X	X	X	X	X	X	X	X	X	X	X	X	S	12
4	X	X	F											2
5	X	X	X	X	X	X	X	X	X	F				9
6	X	X	X	X	X	X	X	X	X	X	X	F		11
7	X	X	X	X	X	X	X	X	X	X	X	X	S	12
8	X	X	X	X	X	X	X	F						7
9	X	X	X	X	X	X	X	F						7
10	X	X	X	X	X	X	X	X	X	X	X	F		11
11	X	X	X	X	X	X	X	X	X	X	X	X	S	12
12	X	X	X	X	X	X	X	X	X	X	X	F		11
13	X	X	X	X	X	X	X	X	X	X	X	F		11
14	X	X	X	X	X	X	X	X	X	X	X	X	S	12
15	X	X	X	X	X	X	X	F						7
16	X	X	X	X	X	X	X	X	X	F				9
17	X	F												1
18	X	X	X	X	X	X	X	X	X	X	X	X	S	12
19	X	X	X	X	X	X	X	X	X	X	F			10
20	X	X	F											2
21	X	X	X	X	X	X	X	X	X	X	X	X	S	12
22	X	X	X	X	X	X	X	X	X	X	X	X	S	12
23	X	X	X	X	X	X	X	X	X	X	F			10
24	X	X	X	X	X	X	X	X	X	X	X	X	S	12
25	X	X	X	X	X	X	X	X	X	X	F			10
26	X	X	X	X	X	X	X	X	X	X	X	X	S	12
27	X	X	X	X	X	X	X	X	X	X	X	X	F	12
28	X	X	X	X	X	X	X	X	X	X	X	X	S	12

	0	1	2	3	4	5	6	7	8	9	10	11	12	[b]
29	X	X	X	X	X	X	X	X	X	X	F			10
30	X	X	X	X	F									4
31	X	X	F											2
32	X	X	X	X	X	X	F							6
33	X	X	X	X	X	X	X	X	X	X	X	X	S	12
34	X	X	X	X	X	X	X	X	F					8
35	X	X	X	X	X	X	X	X	X	X	X	X	S	12
36	X	X	X	X	X	X	X	X	X	X	X	X	S	12
37	X	X	X	X	X	X	X	X	X	F				9
38	X	X	X	X	X	X	X	X	X	F				9
39	X	X	X	F										3
40	X	X	X	X	X	X	X	X	X	X	X	X	S	12
N_x	40	39	36	35	34	34	33	30	29	24	20	16	15	370

Total

	0	1	2	3	4	5	6	7	8	9	10	11	
d_x	0.9750	0.9231	0.9722	0.9714	1.0000	0.9706	0.9091	0.9667	0.8276	0.8333	0.8000	0.9375	
Survival[c]	0.9750	0.9000	0.8750	0.8500	0.8500	0.8250	0.7500	0.7250	0.6000	0.5000	0.4000		0.3750

[a] Day 0 is the day on which the first egg was laid in the nest. X indicates the presence of eggs or chicks in the nest. Young fledge on day 12 (the 13th day of a nest episode). S indicates successful departure of at least one chick. F indicates nest failure (no chicks known to have fledged).

[b] Exposure days for each nest.

[c] Survival is the probability of surviving from day 0 to day x and is given by $\prod_0^x d_x$ ($d_x = N_{x+1}/N_x$). Survival at nest day 0 is 1.00

With the Mayfield method, not all nests are monitored every day. Thus, some nests may fail between one nest check when eggs or chicks are present and the next nest check when they are gone. Mayfield (1975) suggested a method for estimating the number of exposure days that may have occurred during the unobserved interval between nest checks. For the purposes of this analysis, I assume that nests were monitored on every day following nest discovery.

In order to test the efficacy of the Mayfield model, I took a sample from the population shown in Table 1. In the simulation, I pretended that I "discovered" the 40 nests on different days by removing the X's from Table 1 prior to arbitrarily chosen days of discovery (Table 2). For examples: nest 1 was found on the day the first egg was laid and was successful (12 exposure days), nest 2 was found on the day the first egg was laid and failed on day 9 (9 exposure days), nest 7 was found on day 8 and was successful on day 12 (4 exposure days), nest 8 was found on day 4 and failed on day 7 (3 exposure days). Next, I applied the Mayfield method to the data in Table 2 and compared the results with those from Table 1. In the sample shown in Table 2, the number of counted exposure days is 199, and the number of nest failures is 25. Therefore, according to the Mayfield method, DMR is $25/199 = 0.1256$; DSR $= 1 - 0.1256 = 0.8744$; and from Eq. 2, $s = 0.8744^{12} = 0.1997$, which is quite different from the actual nest success of $15/40 = 0.3750$ of this population (shown in Table 1). The Mayfield method, then, does not provide a good estimate of nest success for the actual population from which the sample was taken.

Why should there be so great a discrepancy? I suggest that the Mayfield method makes several serious mathematical errors.

ERRORS IN THE MAYFIELD METHOD

(1) The most serious problem with the Mayfield method is the *exclusion* of exposure days occurring prior to the day of discovery (in Table 2—all spaces before the first X in each row). These arc not counted as days survived. Yet, we can be *absolutely certain* that a nest found any day before fledging had survived from the first day an egg was laid to the day of discovery. We do not have to observe a nest's existence to know that it existed before we found it. Therefore, there is no justification whatsoever for excluding these known days of

exposure. Even if an investigator cannot age the nest and, thus, not be able to determine the exact number of exposure days prior to a nest's discovery, the solution to this problem cannot be to exclude those days altogether.

Suppose we count these excluded days as exposure days. Although these cannot usually be counted in field studies, we can count them in this simulated study. By filling in the missing exposure days in Table 2, we obtain Table 1. Let us calculate nest success of the sample in Table 1 with the Mayfield method. Because we have included every nest on every day of its existence (Table 1), we count 370 exposure days. Mayfield's DMR is $25/370 = 0.0676$; DSR $= 1 - 0.0676 = 0.9324$; and $s = 0.4319$, greater than the observed nest success (0.3750) and more than double that of the Mayfield method (0.1997) with the sample in Table 2. What is the probability of nest success for the population illustrated in Tables 1 and 2: 0.1997, calculated with 199 exposure days (Table 2), or 0.4319, calculated with 370 exposure days (Table 1), both being calculated in the same way with the Mayfield method? Or, is it 0.3750?

These results show us that the Mayfield equation for daily mortality rate (DMR),

$$Daily\ mortality\ rate = \frac{number\ of\ failures}{number\ of\ exposure\ days}, \qquad (3)$$

provides incorrect estimates of the daily mortality rate, which translates into incorrect estimates of nest success. Given a sample of nests, say 100 nests with 55 successes over the study period (i.e., simple nest success is 0.55), nest success is a function of the number of exposure days, which itself is affected by, for example, the frequency of nest censuses and the skill of investigators in finding nests. Assume our study species has a 28-day nest episode from first egg-laying to first fledging. If we counted 500 exposure

TABLE 2. Simulated spreadsheet showing the histories from time of discovery to either success (S) or failure (F) of 40 nests in a hypothetical population. This is a sample of the nests in Table 1.

Nest	Day of Nest Episode (x)[a]													Days[b]
	0	1	2	3	4	5	6	7	8	9	10	11	12	
1	X	X	X	X	X	X	X	X	X	X	X	X	S	12
2	X	X	X	X	X	X	X	X	X	F				9
3									X	X	X	X	S	4
4	X	X	F											2
5									X	F				1
6										X	X	F		2
7									X	X	X	X	S	4
8					X	X	X	F						3
9			X	X	X	X	X	F						5
10	X	X	X	X	X	X	X	X	X	X	X	F		11
11									X	X	X	X	S	4
12								X	X	X	X	F			4
13									X	X	X	F			3
14	X	X	X	X	X	X	X	X	X	X	X	X	S	12
15	X	X	X	X	X	X	X	F						7
16								X	X	X	F			3
17	X	F												1
18										X	X	X	S	3
19	X	X	X	X	X	X	X	X	X	X	F			10
20	X	X	F											2
21						X	X	X	X	X	X	X	S	7
22	X	X	X	X	X	X	X	X	X	X	X	X	S	12
23										X	X	F		2
24										X	X	X	S	3
25									X	X	F			2
26										X	X	X	S	3
27										X	X	X	F	3
28	X	X	X	X	X	X	X	X	X	X	X	X	S	12
29										X	X	F		2

	1	2	3	4	5	6	7	8	9	10	11	12	13	
30	X	X	X	X	F									4
31	X	X	F											2
32		X	X	X	X	F								4
33										X	X	X	S	3
34	X	X	X	X	X	X	X	X	F					8
35										X	X	X	S	3
36											X	X	S	2
37	X	X	X	X	X	X	X	X	X	F				9
38							X	X	X	F				3
39			X	F										1
40	X	X	X	X	X	X	X	X	X	X	X	X	S	12
Total														199

[a] The nest episodes have been truncated to 13 days [twelve 24-hour intervals] in order to save space. The first X in a row is the day of discovery. The subsequent X's indicate the presence of eggs or chicks in the nest. (Exposure days are plotted here with respect to the day of the nest episode only for reference to exposure days in Table 2. Ordinarily, the day in the nest episode on which a nest is discovered is unknown to the field investigator.)

[b] Exposure days for each nest.

days, the Mayfield DMR = 0.090, DSR = 0.910, nest success = 0.071. If we counted 1,000 exposure days, DMR = 0.045, DSR = 0.955, nest success = 0.275. If we counted 2,000 exposure days, DMR = 0.022, DSR = 0.978, nest success = 0.529. What is the nest success of these 100 nests: 0.071, 0.275, or 0.529? Or, is it something else?

Ornithologists do not often report data that would allow us to evaluate the relationship between nest success and counted exposure days. In one study, however, Klett and Johnson (1982) censused a Mallard (*Anas platyrhynchos*) population in North Dakota at weekly intervals. Using their data, they simulated four samples by calculating "hatching rates" (i.e., nest success) as if the population had been censused every one, two, three, or four weeks. As searches decreased in frequency, the number of nests found decreased, the average age of nests at discovery increased, and the Mayfield "hatch rates" increased (23 percent greater when censused every four weeks than when censused at weekly intervals). Is the Mayfield nest success for this population 20.7 (one week search intervals), 23.6 (two week intervals), 24.2 (three week intervals), or 25.4 (four week intervals) (Klett and Johnson, table 2)? Or, is it something else? Again, the Mayfield estimate of nest success is a function of the research protocol, increasing as the number of counted exposure days increases, rather than an indicator of the actual nest success of the population.

In another study, Knutson et al. (2007) provided data on 27 species of birds: nests found, nests successful, number of counted exposure days, length of the nesting cycle (what I have been calling the nest episode), daily survival estimate (daily survival rate), and the interval survival estimate (nest success). Thus, we can make an across-species comparison between nest success and exposure days in the following manner. First, from data in Knutson et al.'s (2007) table 5, calculate apparent nest success (nests successful/nests found). Second, calculate total potentially countable exposure days (number of nests found times length of the nesting episode). Third, calculate the fraction of the total potential exposure days that are in fact counted (i.e., counted exposure days divided by total potential number of exposure days). Fourth, calculate the difference between the apparent nest success and the predicted nest success. Finally, plot the difference between the two measures of nest success against the percent of total exposure days that are counted (Fig. 1). This plot shows that the percent of exposure days counted in a particular study has a profound effect on predicted nest success, the latter becoming more different from the apparent nest success as the percent of counted exposure days decreases.

"Exposure days" is largely a measure of investigator effort—the greater the effort, the larger the number of exposure days. Thus, in the Mayfield method, investigator effort has a great effect on calculated nest success. Because investigator effort cannot be considered a component of daily nest survival or nest success, the Mayfield estimator is meaningless as a measure of nest success.

(2) The second mistake is the assumption that one minus the Mayfield DMR is the daily survival rate (DSR). In the world of numbers, however, the daily survival rate is the number of survivors on day $t + 1$, i.e., S_{t+1}, divided by the number at risk on day t, N_t, (i.e., S_{t+1}/N_t). The daily mortality rate is the number of deaths between day t and $t + 1$, i.e., D_{t+1}, divided by the number at risk on day t, N_t (i.e., D_{t+1}/N_t.). Thus, the daily survival rate plus the daily mortality rate equals one: $(S_{t+1}/N_t) + (1 - (S_{t+1}/N_t)) = 1$.

The number of failures divided by the number of exposure days, however, is the "number of failures per exposure day." This is not the mathematical equivalent of the *daily mortality rate*, and, therefore, one minus Mayfield's daily mortality rate is not the equivalent of the daily survival rate (d_t), that is, S_{t+1}/N_t. Inasmuch as Mayfield's DSR is not d_t, nest success (s) cannot be calculated with Eq. 2. Because the number of failures remains a constant in any sample and the number of exposure days is a variable depending on investigator effort, Mayfield's DMR decreases as the number of exposure days increases, resulting in higher nest survival with increasing number of exposure days.

(3) An assumption of the Mayfield method is that the daily survival rate is a constant, which justifies the use of Eq. 2 for estimating nest success. The daily survival rate, however, cannot be a constant under any circumstances for any period of time (other than a constant 100 percent daily success rate)

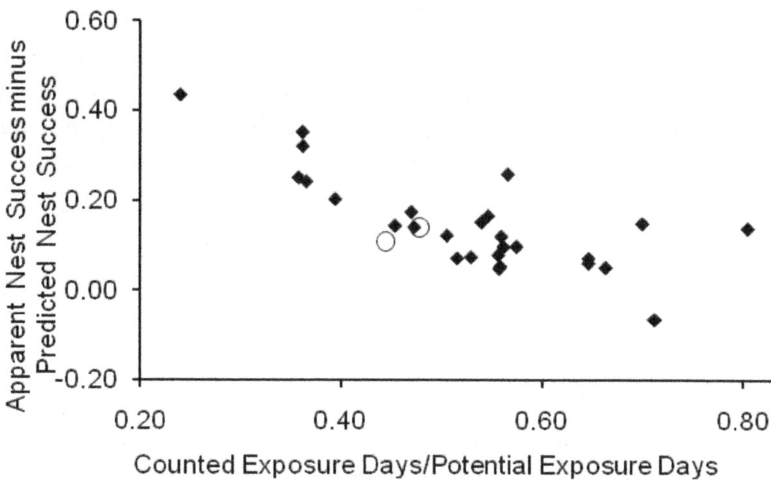

FIGURE 1. The relationship between exposure days and nest success for 27 species of woodpeckers and passerines (data from Knutson et al. 2007) and two species of ducks (Klett and Johnson 1982). The woodpeckers and passerines are represented by diamonds. The ducks are represented by open circles. The "counted exposure days" is the number of days nests are actually observed and counted. "Potential exposure days" is the number of nests found for a species times the length of the nest episode from laying of the first egg to departure of the first fledgling (i.e., the number of exposure days if all nests survived from first egg laying to success). The vertical axis is the difference between the simple apparent nest success and the nest success predicted by the Shaffer model or, in the case of the ducks, the Mayfield method. The difference in nest success decreases with increasing fraction of potential exposure days actually counted.

because N_t and N_{t+1} must always be whole numbers (i.e., nests come in units of one). Suppose that we start with 50 nests and assume that d_t is a constant 0.98, then $N_0 = 50$, N_1

= 49, $N_2 = 0.48.02$, $N_3 = 47.06$, $N_4 = 46.12$, and $N_5 = 45.20$. On each of these days, N_t is not usually a whole number. If we assume that this is merely a rounding problem, we might conclude that the true sequence of N_t is 50, 49, 48, 47, 46, and 45. If so, then the correct daily survival rates are $d_0 = 49/50 = 0.98$, $d_1 = 48/49 = 0.9796$, $d_2 = 47/48 = 0.9792$, $d_3 = 46/45 = 0.9787$, $d_4 = 45/46 = 0.9783$. Clearly, I repeat, the daily success rate cannot be a constant under any circumstances for any period of time (other than a constant 100 percent daily success rate).

Later variations of the Mayfield method claim to relax this assumption, allowing for variations in daily survival rate during the nest episode (e.g., Johnson 1979; Klett and Johnson 1982; Johnson and Shaffer 1990). The "relaxation" of this assumption, however, does not correct for errors 1 and 2.

DISCUSSION

For the reasons given above I conclude that the Mayfield method is completely unsuitable for calculating the probability that a nest will survive an entire nest episode and produce young to fledging. Other methods (e.g., program MARK [Dinsmore et al. 2002; Rotella et al. 2004; Dinsmore and Dinsmore 2007] and logistic-exposure [e.g., Shaffer 2004; Knutson et al. 2007; Shaffer and Thompson 2007]) using "exposure days" in the calculation of nest success should no longer be used. As we have seen, the estimate of the nest success of samples is biased by the age at which nests are discovered, regardless of which method is used. Yet, any method designed to estimate nest success should be independent of the number of days that the nests are found.

In a comparison of methods, Johnson (2007a: 71-72) states, "if an objective [of a study] is to estimate overall nest success, these methods can generate a pooled estimate that is comparable to, say, a Mayfield estimate," and "The Mayfield estimator … has borne out rather well. In a number of comparisons with more sophisticated methods, it has proven competitive." For example, the estimates of the probabilities of nest success for nine populations of ducks with both the Mayfield and logistic-exposure models were "nearly identical," differing by less than 0.01 (Shaffer and Thompson 2007). The estimates of daily survival rates of the Chestnut-collared Longspur (*Calcarius ornatus*), made with the Mayfield method and logistic-exposure (best model), were virtually identical, differing by 0.000 for a population in native habitat and 0.001 in exotic habitat (Lloyd and Tewksbury 2007). Jehle et al. (2004) compared four methods (the simple or apparent method, the Mayfield method, the Stanley method, and program MARK) for calculating nest success for each of six samples of the Lark Bunting (*Calamospiza melanocorys*) in Colorado. The maximum difference between nest success (Mayfield, Stanley, and MARK) in the six samples was less than 0.03, whereas the minimum difference between these estimates and apparent nest success were at least 0.18, 0.01, 0.10, 0.06, 0.23, and 0.12 for these six samples. *If estimates of nest success from these various methods are not substantially different from those of the Mayfield method, it follows that they, too, are mismeasures of nest success.*

The obvious solution to the bias introduced by including late-found nests is to include for analysis only those nests found at or before egg laying starts. Although it is not always possible to find all nests before or at egg laying, we can easily eliminate later-found nests as unsuitable for the analysis, just as, for example, when determining the mean clutch size of a population, we exclude clutches from which eggs have been removed by predators or brood parasites prior to clutch completion.

An important question is whether a sample is representative of the nest success of the population being studied, including all the nests not found. We cannot possibly know the nest success of the total population (including nests not found), any more than we can know the mean weight of the migrant birds flying overhead from a sample of 50 captured in mist nets at ground level. Normally, we accept samples for what they are, samples. We should interpret them judiciously. We should be wary of samples that include many nests found after first laying.

Inasmuch as the simple method (i.e., number of successful clutches divided by total number of clutches) is sufficient for giving an accurate measure of the probability of clutch success for a sample in which the fates (either success or failure) of all nests are known (as illustrated in the example given in Table 1), I recommend that the simple method be used for calculating clutch success with nests found at or before egg laying begins.

ACKNOWLEDGMENT

I thank J. R. Jehl, Jr., for his comments on earlier versions of this paper.

Literature Cited

DINSMORE, S. J., AND J. J. DINSMORE. 2007. Modeling avian nest survival in program MARK. Studies in Avian Biology 34: 73-83.

DINSMORE, S. J., G. C. WHITE, AND F. L. KNOPF. 2002. Advanced techniques for modeling avian nest survival. Ecology 83: 3476-3488.

JEHLE, G., A. A. Y. ADAMS, J. A. SAVIDGE, AND S. K. SKAGEN. 2004. Nest survival estimation: a review of alternatives to the Mayfield estimator. Condor 106: 472-484.

JOHNSON, D. H. 1979. Estimating nest success: the Mayfield method and an alternative. Auk 96:651-661.

JOHNSON, D. H. 2007a. Estimating nest success: a guide to the methods. Studies in Avian Biology 34: 65-72.

JOHNSON, D. H. 2007b. Methods of estimating nest success: an historical tour. Studies in Avian Biology 34: 1-12.

JOHNSON, D. H., and T. L. SHAFFER. 1990. Estimating nest success: when Mayfield wins. Auk 107:595-600.

JOHNSON, D. H. 2007a. Estimating nest success: a guide to the methods. Studies in Avian Biology 34: 65-72.

JOHNSON, D. H. 2007b. Methods of estimating nest success: an historical tour. Studies in Avian Biology 34: 1-12.

KLETT, A. T., AND D. H. JOHNSON. 1982. Variability in nest survival rates and implications to nesting studies. Auk 99:77-87.

KNUTSON, M. G., B. R. GRAY, AND M. S. MEIER. 2007. Comparing the effects of local, landscape, and temporal factors on forest bird nest survival using logistic-exposure models. Studies in Avian Biology 34: 105-116.

LLOYD, J. D., AND J. J. TEWKSBURY. 2007. Analyzing avian nest survival in forests and grasslands: a comparison of the Mayfield and logistic-exposure methods. Studies in Avian Biology 34: 96-104.

MAYFIELD, H. 1960. The Kirtland's Warbler. Cranbrook Institute of Science.

MAYFIELD, H. 1961. Nesting success calculated from exposure. The Wilson Bulletin 73: 255-261.

MAYFIELD, H. F. 1975. Suggestions for calculating nest success. The Wilson Bulletin 87: 456-466.

ROTELLA, J. J., S. J. DINSMORE, AND T. L. SHAFFER. 2004. Modeling nest-survival data: a comparison of recently developed methods that can be implemented in MARK and SAS. Animal Biodiversity and Conservation 27: 187-205.

SHAFFER, T. L. 2004. A unified approach to analyzing nest success. Auk 121: 526-540.

SHAFFER, T. L., AND F. R. THOMPSON, III. 2007. Making meaningful estimates of nest survival with model-based methods. Studies in Avian Biology 34: 84-95.

WEATHERHEAD, P. J. 1989. Sex ratios, host-specific reproductive success, and impact of Brown-headed Cowbirds. Auk 106:358-366.

WEGGLER, M. 2006. Constraints on, and determinants of, the annual number of breeding attempts in the multi-brooded Black Redstart *Phoenicurus ochruros*. Ibis 148:273-284.

Commentary (*Condor*, 24 August 2009)

Dear Dr. ...:

Thank you very much for finding two reviewers who returned their reviews so promptly. My guess is that you found two guys who had reviewed it already for someone else, although their arguments do not seem familiar to me. In any case, I appreciate their promptness. Nevertheless, I am afraid that I will not be able to revise my paper such that you would find it acceptable for publication. You asked for "an itemized report detailing my responses to the reviews." This is what follows. It is long because the reviewers' comments are many. I hope that you will take the time to read my commentary (not necessarily all at once) and, if you have the time, explain to me why I am wrong.

I expect that you will be unhappy with my "tone" in places. I am afraid that my total exasperation with discussing mathematical issues with ornithologists comes through. My responses are not personal (after all, I do not know who the reviewers are; I would write to them more kindly if I did). If the reviewers knew who I was, they probably would have written their comments with more respect for my views.

Before I get started, let me address a few points that the reviewers missed.

(I) I am a biologist, not a statistician. I am interested in the biology of birds. I am really distressed that ecology has become a branch of statistics instead of being treated as scientific natural history. Read any of the recent papers on "nest success," which are loaded with tables of statistics almost to the exclusion of discussing nest success. Where are the tables showing sample size, number of exposure days (or whatever they are measuring), length of nest episodes, and so on? Where is the input data for their models? The reviewers are clearly statisticians who use biology as a source of information for "modeling" what seem to be complex systems. In my view, if you do not get the biology correct, then all the fancy statistics are just a waste of time, energy, and paper. Unfortunately, ornithologists are not getting the biology correct.

(II) I love to crunch numbers. I do not write a sentence with mathematical content without crunching the numbers first. As far as I can tell—from what they write—ornithologists do not crunch numbers. By crunching numbers I mean adding, subtracting, etc. them on a spreadsheet of one's own design. I do not count as "crunching numbers" feeding them into someone else's software and accepting what comes out as correct. I always keep in mind the old adage about using computers, seemingly no longer taught, "garbage in, garbage out."

(III) What has thrown off the reviewers (and perhaps other readers) is that my paper is purely descriptive. I am concerned with what is, not with what might be if I made this or that assumption. I make no *a priori* assumptions, other than that I will use the standard English and the standard arithmetic that I learned in high school or before. I write what I mean, and I do not appreciate others making incorrect inferences that I would not have made and criticizing them as if they represented my views. I will be referring back to this idea later in the commentary.

(IV) I am frustrated throughout the reviews by statements that are made by you and the reviewers that are stated to be true without the writers' providing substantive arguments. For example, you begin by saying, "I trust you will agree that there is no simple 'I think Bert Murray is wrong' in these reviews." You are right, there is no *simple* statement by either reviewer, "I think Bert Murray is wrong." Instead, the reviewers maintain the correctness of their views, without providing any substantive supporting evidence, which is a just another way of saying, "Bert Murray is wrong." Their case would have been stronger if they had offered concrete counterexamples. I will point these out as we go along.

Then you say, "There are passages [in your ms.] that strike me a[s] needlessly combative or bombastic." Wow! That is quite a charge. I am unaware of such passages, and Joe Jehl, who reads my stuff, does not usually allow me to vent my true feelings. It would have been helpful for me if you had pointed out some of those passages. On the other hand, it is difficult to write criticism for an audience as thin-skinned as ornithologists. Again, some identified passages of what you consider offensive would be very helpful.

This is a failing throughout the two reviews. Both reviewers simply make declarative statements as if they were fact. We are supposed to believe their statements without question. As a scientist I do not believe them without question. I will point these out as we go along.

You point out several observations about statistics and say, "it seems apparent to me that to a large degree we are on the turf of philosophy of science and philosophy of statistics. Perhaps, then, your ms. would benefit by making it clear that one's philosophical position will play a role in how these problems are approached." As I pointed out before, my paper is purely descriptive. Whether the Mayfield method or the other methods produce correct estimates of clutch success or not will not be decided by one's philosophy. This determination will be made mathematically (i.e., by numbers crunching). I provided an explicit example in my paper (Tables 1 and 2 and discussion of them). My critics have not.

(V) Before I become too critical of the reviewers, let me point out an issue on which I am in total agreement. The second reviewer wrote, "I don't see the need for the structure of this commentary to mimic a research study (i.e., intro, methods, results, discussion). Right now, the treatment of the simulations is very jumbled across sections. I'd suggest a re-organization …." I totally agree. (By the way, I consider this a "research study," even if it is not an "experimental study.")

The reviewer, however, has not read the Instructions for Authors. Under the heading, Manuscript Order, I am informed, "Correct sequence for sections of a submitted manuscript is Title page, Abstract, Key words, Introduction, Methods, Results, Discussion, Acknowledgements, Literature Cited, Figure legends, Figures, and Tables. ... Each main heading is capitalized (INTRODUCTION, METHODS, RESULTS, DISCUSSION, ACKNOWLEDGMENTS, LITERATURE CITED)." That seems fairly explicit. Furthermore, in the section General Guidelines, I am informed, "Papers that are not in *Condor* format will be returned without review." That seems fairly explicit. That seems commanding. Furthermore, I have never won an argument about style with an editor, even when I was right.

The initial submissions of the ms, by the way, were in a form similar to the one recommended by the reviewer (note: I discuss the second reviewer first).

The second reviewer

Let me start with a few examples. In my letter of submission I suggested that the reviewers directly respond to five questions. You evidently sent the letter to the reviewers because one (the second) actually responded directly. Let us begin with him. He made nine points.

Point 1. My statement, "[T]here is no justification whatsoever for excluding these known days of exposure [the days prior to nest discovery]."

The second reviewer suggests that determining the number of days survived before discovery "requires making a series of assumptions [which are not always solid]. So, we are left with a choice (if we want to use exposure days) of sticking with what we can quantify (days a nest was actually observed) or making our best guess (days a nest might have been active). If our best guess is likely to be wrong (i.e., too much uncertainty in assumptions), then the prudent decision (or justification, if you will) is to not make it."

This is true. I agree. Do not use nests for which you must guess at the number of days the nests survived prior to discovery. This does not mean, however, that you are justified in using nests and a method that are *certain* to give you wrong estimates of nest success. What I should do, as I advised in my ms., is use the nests that have been found at or before egg-laying started. One could probably get away with counting nests found during the first few days of egg-laying, but that would have to be determined ad hoc for each situation (e.g., if no nests were lost during the laying of first, second, and third eggs, one should be able to use nests found with three eggs in them).

It should be remembered that I am discussing the Mayfield method. Thus, I have no choice but to "want" to use exposure days when discussing that method. The Mayfield method is based on determining the "daily mortality rate" (DMR) of the clutches. Mayfield states an explicit equation for calculating the DMR (i.e., DMR = number of failures/number of exposure days). Unfortunately, Mayfield's DMR is not the real world DMR of the clutches. The actual DMR is given by number of failures between day x and day x + 1 divided by number of nests at risk on day x. (See Point 5 below.) Although the number of days a nest was actually observed is countable, these numbers cannot be used to determine the daily survival rate (= 1 − mortality rate) for any day during the nest episode. Thus, sticking with what we can quantify is not an option if you get wrong answers. The second option of guessing the number of days before discovery is not much better than the first. The best option is to limit analysis to the nests found before egg laying starts or very shortly thereafter.

I have not put this next argument in my paper. It is again obvious to anyone with numbers sense. In the real world of mathematics the mortality rate is the number of deaths divided by the number at risk over some period of time, often a day or a year. The survival rate is the number of survivors divided by the number at risk over that same period of time. The mortality rate plus the survival rate over that period of time sums to 1. This is not true in ornithology (e.g., the Mayfield method). Don't you find that just a little bizarre?

Ornithologists are not interested in testing their *belief* that their methods provide correct estimates of daily survival rate. They prefer to assume that "Bert Murray is wrong."

The reviewer continues, "By advocating as you do in the commentary that we ignore all nests not found prior to egg laying, you are advocating that we throw away data by excluding known days of activity from nests found later in the nesting cycle. It wasn't clear to me in reading your commentary why you feel it is appropriate to ignore data in one context and not another. This may not be what you are advocating but I was struck by the apparent inconsistency."

Remember, I am criticizing the Mayfield method. With regard to calculating daily survival rate (in the real world of mathematics), an investigator needs the number of nests at risk on day x and the number of surviving nests on day x + 1. The calculation of nest success, then, requires knowing the number of successful nests day by day throughout a nesting episode. Nests for which data are missing must be excluded. You would not include a dead headless Cardinal in your backyard if you wanted to know the mean weight of the Cardinals in your backyard. You must have intact, preferably living, Cardinals. In the same way, you must have all the days between the laying of the first egg and success or failure to calculate a correct estimate of success for a group of nests. You do not use incomplete samples (nests found after egg laying) because you cannot calculate correct estimates of success. You need nests found prior to first-egg-laying in order to estimate clutch success correctly.

I am not being inconsistent at all.

Point 2. I wrote, "Is the Mayfield nest success for this population [of Mallards] 20.7(one week search intervals), 23.6 (two week intervals), 24.2 (three week intervals), or 25.4 (four week intervals) (Klett and Johnson, table 2)? Or, is it something else?"

The reviewer is not surprised by Klett and Johnson's results. He thinks that, "The variation in Mayfield estimates is, as Klett and Johnson state, likely due to variation in daily mortality rates." That is not a rebuttal of my point because that is an ad hoc saving hypothesis for which they have no evidence. Furthermore, the four "samples" span the identical period of time because there is just one set of nests. If there were any variation in daily survival rates, it should be the same for each sampling method. Klett and Johnson subsampled their single sample.

Remember, we are (or at least I am) interested in determining the survival rate of nests in a population. Of 50 nests, how many can we expect to see survive to fledging their first young? What I want is a method that will give me that probability. That method should give me the same number every time, regardless of the effort that I expend in searching for and monitoring nests. The Klett and Johnson example shows pretty clearly that estimates using the Mayfield method are different depending on the effort expended.

Remember, I must repeat, the Klett and Johnson data are subsamples from a single sample of Mallard nests. Why should the daily mortality rate for these nests differ depending on whether you sample at one-, two-, three-, or four-week intervals?

I see no reason for eliminating this example from my paper. The Mayfield method does not work, which is the point of my paper. I know that is embarrassing for the promoters of the Mayfield method or its alternatives, but that is too bad— that is, if ornithology is a science.

Point 3. I wrote, "This plot shows that the percent of exposure days counted in a particular study has a profound effect on predicted nest success, the latter becoming more different from the apparent nest success as the percent of exposure days counted decreases."

(3A) The reviewer comments, "I'm not convinced that the denominator you have chosen (# of exposure days if all nests were successful) is relevant as we know that not all nests were successful. The comparison of what was observed to something we know not to be true doesn't serve much purpose. A more relevant denominator (and, unfortunately, something that we can't get from the data presented in Knutson et al.) would be the number of exposure days available to be counted if (a) all nests were found prior to egg-laying and (b) we had the field staff available to monitor all nests daily. This would be a much better index of your central point regarding the role of investigator effort and number of exposure days."

In fact, I did not determine the denominator as the reviewer thinks. The denominator is *not* the "# of exposure days if all nests were successful," as he says. You will not find that definition in my paper. In fact I determined the "much better index," as the reviewer calls it, that is, "the number of exposure days available to be counted." In fact, contrary to the reviewer's thinking, Knutson et al. did provide the data in their Table 5 for me to do just that. They report sample size (number of nests found) and the length of the "nesting cycle," which I presume is the time between the laying of the first egg and first fledging. This should be "the same" for all nests. Thus, in the case of the redstart, the number of exposure days available to be counted is 196 nests X 24 days = 4,704 potential exposure days. The reviewer's misreading of my method does not refute my argument.

(3B) Here comes the statistical argument. The old reliable criticism. If reason does not work, try statistics. Does the reviewer really believe that the result of the statistical test will show that the data in Figure 1 have a different shape? Suppose that a statistical test shows that the relationship is statistically not significant from—what?—a straight line? Suppose that the statistical test shows that the observed curve is not statistically different from a straight line. Isn't there the danger of rejecting a true relationship?

In any case, the Mayfield equations actually *predict* a positive relationship between nest success and number of exposure days (however counted)—see next figure. To know this, one must crunch some numbers without dreaming up a dozen or so ad hoc alternative and unsubstantiatable hypotheses.

The *inference* from an analysis of the empirical data on several species, plotted in Figure 1 of the manuscript [pg. 201], is that nest success increases as the number of counted exposure days increases. The *prediction* of the Mayfield equations is that nest success increases as the number of exposure days increases (see figure on next page).

I was tempted to discuss this issue and figure in my paper but decided not to because I thought I would be beating a dead horse. No doubt the statistician would demand statistics of some sort. Prove the obvious with statistics. Ridiculous.

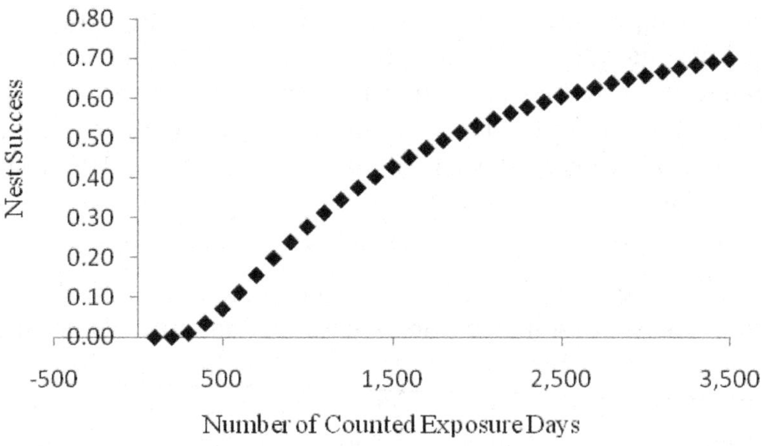

Legend for above figure: Prediction of nest success using Mayfield equations.

(3C(a)) The reviewer writes, "I'm confused by your discussion regarding the Hairy Woodpecker." He continues, "Table 2 in Knutson et al. (at least the version I have) shows that 21 of 21 Hairy Woodpecker nests were successful, not the 14 of 21 you note in your commentary. You mention a personal communication with Melinda; did she indicate that the data in her published manuscript were incorrect?" The short obvious answer is, "yes." The longer answer is that an earlier reviewer thought that 21 successes out of 21 nests was unbelievable and checked it out with Knutson. He found out that it was 14 out of 21. In his review, he condescendingly pointed out to me that I could have found that out for myself by writing her. I fact, I had written to her about the Hairy Woodpecker data, but she did not reply (typical for an ornithologist). When I wrote to her for her to confirm the reviewer's numbers, she replied that she thought she had written to me. I assumed that she was being honest and that the mail went astray. I thought my "in litt." reference to Knutson for the correction and my acknowledgement to her ("I thank M. G. Knutson for clarifying data in her paper (Knutson et al. 2007)") would have been sufficient for a reader to understand that she had confirmed the 14 of 21. Now, how much of this story am I supposed to include in my paper?

(3C(b)) The reviewer writes, "The interval success for the Hairy Woodpecker is not 'by far the lowest nest success of the 27 species.' There is one species with lower interval nest success (Song Sparrow at 0.178) and another that is close (Northern Cardinal at 0.246)." The reviewer is correct. I will have to change the wording.

(3C(c)) The reviewer writes, "'Can it possibly be because the sample contained the fewest exposure days of all the samples?' No. There are eight species with fewer exposure days that [sic] the Hairy Woodpecker." Right again! What I meant to write was, "Can it possibly be because the sample contained the smallest fraction of the potential number of exposure days that could be counted?"

(3C(d)) The reviewer writes, "You're right to call-out the striking difference between apparent success and interval success for this species, a difference that is even more apparent when you consider that 21 of 21 were successful. However, I'm not sure this example is as illustrative as you think it is; as I note below, in my experience most studies of cavity-nesting species underestimate the number of unsuccessful nests (largely due to the difficulty in checking nest contents), thereby biasing all nest success estimators." Wow! Whose side is he on in this argument?

I don't get it. It's OK for Knutson et al. to publish data for species in which studies, according to the reviewer, "underestimate the number of unsuccessful nests (largely due to the difficulty in checking nest contents), thereby biasing all nest success estimators," but I am chastised for using these same data as examples of incorrectly estimated nest success. I do not see any caveats about the quality of the estimates in the Knutson et al. paper.

(3D) I don't get it. (Am I repeating myself?) The reviewer writes, "Five of the 6 that do not [fall within the CI of the interval nest success estimated using the Shaffer method] are cavity-nesters, a guild for which it is very difficult to locate unsuccessful nests, thereby biasing all nest success estimators, including the simple method. If you wish to retain this section of your commentary, I would suggest removing cavity nesters from the analysis." Say what?! In my paper criticizing the methods of calculating nest success, I am supposed to leave out published data, which were good enough to be published in the first place, because they were not obtained from species good for the method I am criticizing? Shouldn't the authors of that paper have pointed this out? I really do not get it.

(3E) The reviewer writes, "Knutson et al. (2007) do not mention constraining their analyses to just those nests found at or before egg-laying. By your own definition, this renders any estimate of nest success (including the simple method) inaccurate and, as a result, calls into question the validity of your response variable." Again, extraordinary.

Point 4. The reviewer questions my statement: "'Exposure days' is largely a measure of investigator effort—the greater the effort, the larger the number of exposure days." He makes two points. Regarding the first point, the reviewer is correct (but it is a matter of definition), which was undiscovered by the dozen or so previous reviewers and editors who have read it.

The reviewer's second point is too ad hoc to think in detail about. He makes an argument based on the assumption that the Mayfield method incorrectly estimates the number of exposure days. I did not make that point, and I probably won't, but I won't challenge it. Doesn't it further support my point that Mayfield method is flawed?

The reviewer states, "The main ways that I can envision an increase in investigator effort leading to an increase in exposure days is through finding more nests and finding more nests earlier in the nesting cycle (something you advocate elsewhere in the commentary)." Instead of reading my paper as if I were writing what I meant, and even though I had advocated the point earlier in my paper, he invents an off-the-wall way of reading this section.

Point 5. The reviewer quotes me, "This is not the mathematical equivalent of the *daily mortality rate*, and, therefore, one minus Mayfield's daily mortality rate is not the equivalent of the daily survival rate (d_x), that is, N_{x+1}/N_x."

The reviewer comments, "I'll admit that I don't quite get what you're worried about here. The Mayfield method generates, as you note, an estimate of the number of nests that fail while being observed (number of exposure days). This estimate is then assumed to be indicative (as any sample might be) of the daily mortality rate experienced by the population. It strikes me that your primary issue is one of semantics (but I beg your indulgence if I've misinterpreted what you're concerned about)."

I am sorry, but I am not going to indulge your misinterpretation. First, the Mayfield method does not "estimate" the number of nests that fail while being observed (number of exposure days). The Mayfield method *counts* the number of failures and divides that number by an *estimate* of the number of exposure days.

The point was and the fact remains that, mathematically, the number of failures divided by the number of exposure days is NOT an indicator or an estimate of the daily mortality rate—period. Ornithologists must crunch some numbers if they really want to know this, but ornithologists would rather believe in untruths than crunch numbers (I beg you to indulge me my sarcasm here, but that's the way it looks to me).

Point 6. The reviewer quotes me, "An assumption of the Mayfield method is that the daily survival rate is a constant, which justifies the use of Eq. 2 for estimating nest success." This, by the way is a true statement.

Even though I wrote in the next paragraph, "Later variations of the Mayfield method claim to relax this assumption, allowing for variations in daily survival rate during the nest episode (e.g., Johnson 1979; Klett and Johnson 1982; Johnson and Shaffer 1990). The 'relaxation' of this assumption, however, does not correct for errors 1 and 2," the reviewer takes me to task for having even mentioned it. But remember, I am criticizing the Mayfield method.

The reason I went into so much detail in the previous paragraph is that I believe, from what they write, that ornithologists do not understand that daily survival rates cannot be constant—ever (that is, *ever*). Not in this species, not in that species, not in this model, not in that model, *not ever*. I believe ornithologists do not understand this, much less why it should be.

Point 7. The reviewer quotes me, "If estimates of nest success from these various methods are not substantially different from those of the Mayfield method, it follows that they, too, are mismeasures of nest success."

The reviewer comments, "This statement is internally logically consistent if you agree that Mayfield estimates are mismeasures of nest success. You have highlighted what you feel to be important issues with the Mayfield estimator but have not convinced me that (a) your preferred method is any more indicative of the true population or (b) that your preferred method is any less prone to bias, albeit different forms than the Mayfield estimate."

First (a), we can never know what the "true" nest success is in a population because normally we can never find all of the nests, just as we cannot know the "true" mean weight of individuals in a population because we normally cannot weigh every individual. We are stuck with having to rely on samples. Second (b), my preferred method of using only those nests found at or before the laying of the first egg is the best of the available sampling methods. Even critics of the "simple" method agree with that. The problem for the statisticians among us is the fact that, in many studies, not all of the nests can be found prior to first egg-laying. So, they go to work, ignoring the mathematics, designing spurious models to make the estimate. It is not for them to suggest that ornithologists do a better job of finding nests. Some field ornithologists, however, seem able to do this.

It would have been helpful if the reviewer had pointed out the biases of my preferred method. His criticism, as it stands, is another belief without foundation.

Point 8. The reviewer raises several points with regard to my statement, "Although it is not always logistically possible to find all nests before or at egg laying, we can easily eliminate later found nests as unsuitable for analysis."

Reviewer's Point A. See my Point 1 above.

Reviewer's Point B.

The reviewer comments, "How do you (or anyone for that matter) define 'later found'? All nests not found prior to egg laying? All nests found during incubation? You spend a great deal of time highlighting what you feel is a significant problem but spend almost none providing a viable solution (other than the statement above that advocates ignoring data)."

I think the proper grammatical interpretation of my sentence quoted at the beginning of Point 8, is that "later found nests" are those found later than the day of first egg laying. How else can you read this?

The reviewer claims that I spent almost no time providing a viable solution. At times I know that he has read my paper, but at other times I am not so certain. Remember, my paper is about showing the mathematical flaws in the Mayfield method. The Mayfield method is flawed because it counts exposure days, including the exposure days of nests that were found after first egg-laying. The Mayfield method is flawed even if the investigator has found all the nests from day of first egg-laying (as I showed in my paper). Nests found after first egg-laying *are not suitable* for calculating clutch success. Why is the reviewer so dismissive of "my" solution (proposed by others) of excluding unsuitable data from the analysis?

Reviewer's Point C.

The reviewer claims that I "appear to be advocating that we only study species that are easy to study." I am not advocating that at all. I am advocating that we study any species for which we are able to obtain the data necessary to study whatever problem we are interested in studying. Easily studied species are easier to study than species that can be studied with difficulty. Remember the first rule of selecting a thesis project: pick a problem that will produce results in less than four years.

The reviewer's comment does not make my arguments about the Mayfield method wrong, whatever he thinks that I am advocating (again without evidence).

Point D. I wrote, "It should be noted that, although difficult to do, finding a sufficient number of nests early in the nest episode can be done (e.g., Nolan 1978; Clark and Robertson 1981; Weatherhead 1989; Weggler 2006)." The reviewer claims that he is "not sure" that "all of the four studies you cite … provide support for your statement or provide useful information for most practitioners."

Point D(a). The reviewer writes (I am sorry to quote the reviewer so much, but it seems necessary for readers to follow the arguments), "In Chapter 33, Nolan (1978) presents nest success analysis of 2 samples of nests. In the first sample (15 years), Nolan estimated that 85% were found during active building and all found nests were included in the analysis. In the second sample (9 years), all nests were found during nest building. Estimates of nest success did not differ between the two samples, despite the difference in nest inclusion rules. Also, I'm not sure that presenting a 9-year and 15-year study as a benchmark for setting appropriate sample sizes for analysis is very useful or practical, despite the fact that we all wish we could develop these types of long-term data sets (or could receive financial support to do so)."

First, why does the reviewer not learn from this that one can make good estimates of nest success of the Prairie Warbler even when as many as 15% of nests were found after active nest building? When were those 15% found? During egg-laying or in the days just prior to fledging? (Some failed nests were included in one sample.) Should it be noted that he did not use the simple method—not the Mayfield method—to calculate nest success?

Second, remember my point: despite investigators' justifying their incorrect methods for calculating nest success on the grounds that it requires them to look for nests every day rather than less frequently, some ornithologists are able to find most of their nests during the nest building and early egg laying stages. Nolan is one who did so. I was certainly not advocating that investigators study their birds for 9 or 15 years. If investigators can obtain sufficient data in one year, that is fine. Whether or not I was, this does not show that my analysis of the Mayfield method is in any way wrong. (Critics really ought to stick to what I have in fact written, rather than invent inferences that I do not subscribe to and then "criticize" me for them.)

Why does the reviewer think that he is justified to infer that I am advocating that everyone do a 9- or 15-year study in order to determine nest success, just because those are the length of studies in the literature I cited in a different context? Remember my paper is about showing the flaws in the Mayfield method for calculating clutch success.

Point D(b). The reviewer wrote, "Clark and Robertson (1981) did, in fact, restrict analysis to only those nests for which the data of clutch initiation." Yes.

Point D(c). The reviewer wrote, "Weatherhead (1989:359) states that "most nests were found before egg laying" but I couldn't find a mention of specific sample sizes. Was most 51% or some larger number?" When the reviewer is right, the reviewer is right. In the context of ornithological writing, I took "most" as meaning "most," which is substantially greater than 51%. I can delete this reference if you wish. This does not make my criticisms of the Mayfield method wrong.

Point D(d). The reviewer writes, "Weggler (2006) only found 47% of his nests prior to clutch initiation and did not use when a nest was found to constrain entry into his analysis. As with Nolan, I'm not sure using a 10-year study is a useful benchmark for most practitioners. I also find it interesting that you chose to cite a study that utilizes the Mayfield method." Weggler (2006:275) wrote, "Forty-seven per cent of nests were found before clutch initiation, 23% during the egg stage, ..." I took this to mean that 70% of nests were found before or during the first few days of nest life. Evidently, I see now, Weggler's "egg stage" includes the incubation period. I'll have to delete this reference.

I see no reason for not citing a paper in which the author (I thought) found most nests during the first five days of nest life and used the Mayfield method for analysis.

Point 9. Entitled "Somewhat random thoughts."

A. Potential issues with the simple method.

Please remember that I do not advocate using the simple method unrestrictedly. I advocate that we use the simple method on nests that have been found prior to the laying of the first egg. I will accept in certain cases nests found during the egg laying stage, decided on an ad hoc basis (mentioned above).

(a) The reviewer asks, "What suggestions can you offer to help make samples as representative [of the population] as possible?" We can *never* know whether our samples are representative of the true population—unless, of course, we have data on virtually every individual in the population. In my statistical world (please don't tell me that I am contradicting myself—I have taken a course in statistics), I would increase the sample size (one of the early rules that I learned in my course) by trying to obtain data on as many nests as I could find prior to the laying of the first egg. As the final sentence of my paper states, "I recommend that the simple method be used for calculating clutch success with nests found at or before egg laying begins." I just don't think it is legitimate for the reviewer to intimate that I am recommending that ornithologists use the "simple method" that Mayfield was criticizing.

(b) The reviewer writes, "Biased toward successful nests, which are more likely to be found than unsuccessful nests; this is slightly different than the observation (made by you and Mayfield) that later-found nests are more likely to be successful than earlier-found nests. How do you account for this bias towards overestimating nest success using the simple method?"

I account for this bias by eliminating it. I limit analysis to those nests found prior to first-egg-laying or very early in the egg-laying stage. Thus, there is no bias with respect of their later success or failure. I am not "missing" nests that have failed prior to discovery.

(c) The reviewer writes, "Is the percentage of nests that are successful (the measure you are proposing) the same as the probability that a single nest is successful (the measure you say you are interested in)? I just flipped a coin 20 times and it landed on heads 14 times, resulting in a success (of landing on heads) percentage of 70%, which is different than the probability of a single flip landing on heads (50%)."

I can at least begin to wonder whether the reviewer has taken a course in statistics. If you were betting with someone on which side of your coin would turn up in the good ol' days of Dodge City, you would eventually be shot. I don't think that the guys you scammed would want to hear your lecture on statistics and probabilities. When a coin turns up heads 70% of the time, we usually conclude that it is severely out of balance and play with a different coin, one that over the long run turns up heads on 50% of the flips. Well, maybe 20 tosses was not long enough to get a good estimate of the probability of heads on your coin, but by then you would be dead. I am not going to look up the statistics on coin tossing. It is really beside the point.

With regard to nest success, the answer is, "yes," "the percentage of nests that [is] successful … [is] the same as the probability that a single nest is successful." That's how one calculates "individual" probabilities (that is, if you crunch numbers). I just cannot begin to guess what a computer might say, especially after reading this review.

B. The reviewer writes, "This is not directly germane to this conversation, but I am left wondering whether your simple measure of nest success is a useful measure for someone interested in life history evolution. Perhaps I'm missing something here, but my sense is that a measure such as pair success (i.e., the percentage of breeding pairs in a population that raise a successful brood) might be more informative. Thinking to my own field work, I recall a particularly bat [sic—I cannot guess what the word might be] in which only 9% (4 of 45) of nests fledged at least 1 young; however, we were only monitoring 20 breeding pairs that year, which means that 20% of the breeding population successfully fledged young (largely due to ability/willingness to re-nest after failure). Which of these estimates of success is most relevant?"

I think I can safely say that I think a person interested in life history evolution would find the restricted simple method useful. I am rather surprised that the reviewer should have asked this question of, of all people, me. I have written extensively on life history evolution (Murray 1979, 1984, 1985a, 1985b, 1990, 1991a, 1991b, 1992a, 1992b, 1994a, 1995a, 1999a, 1999b, 2000a, 2000b, 2001, 2006, Jehl and Murray 1986, 1989, Murray and Nolan 1989, 2007, Murray, Fitzpatrick, and Woolfenden 1989) and on population dynamics (Murray 1979, 1982, 1985c, 1987, 1988, 1990, 1991c, 1994b, 1995b, 1997a, 1997b, 1999c, 1999d, 2000c, 2000d, 2003, 2005, Murray and Gårding 1984). I have proposed several new equations, including the clutch size equation (Murray and Nolan 1989),

$$C = \frac{a+1}{\sum_{\alpha}^{\omega} \lambda_x \sum_{1}^{n} P_i},$$

and an equation for calculating Annual Reproductive Success (number of fledglings reared per year) (Murray 1991),

$$\text{ARS}(k) = \sum_1^n P_i k_i = \sum_1^n c_i s_i k_i = c_1 s_1 k_1 + c_2 s_2 k_2 + \ldots + c_n s_n k_n.$$

See papers for definitions, except for s_i, which is the probability of the s_i^{th} clutch producing at least one fledgling from time of first-egg-laying, that is, clutch success.

Among my unpublished papers (www.21st-century.org/bertrammurray.htm) is my life history equation,

$$\frac{a+1}{\sum_\alpha^\omega \lambda_x \sum_1^n P_i}(\textit{for males}) = C = \frac{a+1}{\sum_\alpha^\omega \lambda_x \sum_1^n P_i}(\textit{for females}),$$

with which one can successfully predict (genuinely predict) empirical facts about the life histories of birds and other animals (e.g., clutch size, mating systems). You can see that clutch success is in every one of them. Pair success is not. Neither is clutch size, for that matter.

The reviewer speculates about two measures of success—the fraction of nests that produce at least one fledgling and the fraction of pairs that breed successfully. He asks, "Which of these estimates of success is most relevant?" I must ask, relevant to what? For what it is worth, I have never seen anyone measure the number of pairs that were successful for any purpose. I am afraid that the reviewer must be more explicit. In population dynamics, ecologists usually analyze contributions to population growth one sex at a time. Analysis of either sex should give the same answer (that is mathematically demonstrated in one of my papers).

I think the reviewer has missed something.

The first reviewer

I am not convinced that this reviewer understood *any* aspect of my paper.

(1) Except that I can't disagree with him here.

(2) The reviewer states, "I think that the paper focuses too much on the Mayfield method and not enough on maximum-likelihood estimators that now supersede the method and allow for greater flexibility."

First, I stated in my paper, "Given the diversity of models and applications, I emphasize that I am concerned in this paper *only* with what Johnson (2007a) called the most important question—what is the probability that a nest in a sample of nests will survive a full nest episode from the laying of the first egg to the departure of the first fledgling from the nest." Maximum-likelihood estimators may have superseded the Mayfield estimates in the reviewer's mind, but they have not in the minds of many other ornithologists. If maximum-likelihood estimators are accepted by some, it has been a political choice because they have not demonstrated satisfactorily to anyone but the believers that the Mayfield estimates are unacceptable. The Mayfield method may be unacceptable for doing what other things later models were allegedly designed to do, such as estimate daily survival rates, but I have not seen anyone claim that it is unacceptable for estimating nest success. Indeed, maximum-likelihood estimates of nest success do not seem to differ much from Mayfield estimates.

The fact that others prefer maximum-likelihood estimators to the Mayfield method does not constitute evidence against my arguments about the mathematical flaws of the Mayfield method.

Second, the reviewer states, "If, for example, one uses models that allow for time-variation in daily survival rates, the results for daily survival rates are exactly what can be obtained from the raw data shown in Table 2 of the manuscript." He has to be kidding me. The reviewer is not clear here, but see below.

Third, the reviewer writes, "Likelihood approaches (which have been in existence for many years) don't require any mid-point assumptions or calculation of exposure days, again something that the author regards as a problem with the Mayfield method." As I read my paper, I didn't offer an opinion on the mid-point assumptions (I presume regarding intervals between last observation of live birds in the nest and first observation of failure). What I did was write, "For the purposes of this analysis, I assume that nests were monitored on every day following nest discovery." In this way, I thought I could get around another complicating factor that has no relevance to the essential issue—that the Mayfield method is mathematically flawed. To be explicit, I don't care about estimating the time of failure of nests in samples of incomplete data. The fact that I skip this point does not constitute evidence that my arguments about the flaws of the Mayfield method are wrong.

The reviewer writes, "The author focuses on the advantages of using the data from time of nest-initiation to time of discovery. Methods described in papers by Heisey and by Nur do work with time-to-event analyses, are certainly preferred by some, and would be useful to present."

The reviewer does not grasp what my paper is about. First, I am not writing a review paper on the various methods of estimating clutch success. As I stated in my paper (I am afraid that I must repeat), "Given the diversity of models and applications, I emphasize that I am concerned in this paper *only* with what Johnson (2007a) called the most important question—what is the probability that a nest in a sample of nests will survive a full nest episode from the laying of the first egg to the departure of the first fledgling from the nest."

All these side issues brought up by the reviewers, that their beliefs are correct—do not constitute evidence that I am wrong in what I have written.

My paper is supposed to show the flaws in the Mayfield method. And I think I did this well. That others have created methods that somehow take account of the days between first-egg-laying and discovery is completely irrelevant. Even if Nur et al. did do this, it should not go unnoticed that the Mayfield estimate and the Kaplan-Meier estimate, given by Nur et al., are virtually identical (all years, 0.405 vs. 0.398, respectively). What is the point? The fact that these two measures are "the same" does not refute my points that the Mayfield method provides false estimates and, therefore, so does the Kaplan-Meier estimate.

Heisey et al. (2007) is the perfect example. Despite its title, "The ABCs of nest survival: theory and application from a biostatistical perspective," the paper is not really about nest success. It is about statistics, as a cursory perusal would show to anyone. The authors present or discuss several statistical models and several equations, but nowhere could I find a simple statement of the nest success of the Blue-winged Teal, the subject of their paper. Figure 2 presents the "Raw data for 216 Blue-winged Teal nests," and Figure 3 presents "Estimated survival curves." But what is the clutch success of the Blue-winged Teal? I could not find it. The vertical axis of Figure 3 is labeled "Survival" and plotted against age of the nest in days, but what kind of survival? Nest survival?—No. The graph makes sense only if one assumes that "survival" means "daily survival rate." What this tells me is that the authors really do not know what they are doing, much less what the nature of the problem is. Mind you, the statistics may be great, but the biology is incredibly poor. Whatever they are up to—whether the message, whatever it might be, is right or wrong—does not show that the Mayfield method is right and that I am wrong.

Fourth, the reviewer offers his opinion, "I think it is

inappropriate to conclude that a variety of published methods should no longer be used when a number of those methods have been carefully tested and shown to provide very good estimates under the assumptions that accompany the methods."

That's nice, but my opinion is that these methods have no more been carefully tested than the Mayfield method, which has extraordinarily elementary mathematical errors, as I have shown, and which the reviewer seems not to address. He says, "it is not reasonable to conclude that the methods don't work." I disagree, it is perfectly reasonable if those methods produce incorrect answers to the question.

Fifth (a), the reviewer states, "Daily survival rates are not constant in Table 1 [of my paper]. It is not clear how the data were generated or why but that would be useful information to provide." I guess the reviewer did not read my paper. I wrote,

"if we have been successful in following each of the 40 nests daily from the day the first egg was laid until nest failure or success (at least one fledgling leaves the nest), then we could calculate the daily survival rate of nests (d_x), that is, the number of nests that survive from day x to day $x + 1$ divided by the number of nests at risk on day x (i.e., $d_x = N_{x+1}/N_x$). With these data, we could determine the probability of clutch success (s) with,

$$s = \prod_0^{f-1} d_x ,$$ (1)

where f is the first day on which at least one young leaves the nest, and the first egg is laid on day 0. With this equation, d_x may vary from day to day. More explicitly, from an imaginary set of nest-record cards, I entered on a spreadsheet (Table1) the day the nest was discovered (the first X in each row in Table 1), the subsequent days when eggs or young were found (also marked X), and the day of success (marked S) or failure (marked F)."

Can anyone seriously doubt that I made up my simulation out of whole cloth? Can anyone have any doubt how Table 1 was constructed?

The reviewer continues, "Is the author trying to evaluate how it performs when survival varies through time." Well, no. Inasmuch as I have not even mentioned Mayfield or the Mayfield method up to this point in the paper or even the word "test" in any context, I could hardly be testing the Mayfield method or, for that matter, testing anything. What I am showing here is how the daily survival rate and nest success are calculated when the investigator has all the data, which is something that the reviewer seems unfamiliar with. If he cannot calculate daily survival rate and nest success when he has all the data in a simulation, why should I be confident in anything that he says about anything on this subject?

Fifth (b), the reviewer writes, "When the data are altered for Table 2, how was the deletion of data achieved and why was only a single deletion obtained? This seems important to understand because one is now sampling from the data and we only observe a single sample." Boy, I have no idea what the reviewer is writing about, but I will explain what is going on. Table 1 represents the population that we are sampling. As investigators, however, we have *no idea* of what a field population would look like. That is why it has to be simulated. Table 1 provides *no* information to the investigator—*none*—so it is wrong to refer to Table 1 as "data." We sample a population by going into the field and looking for and monitoring nests. I do not know what "deletion of data" means, but I did explain, pretty clearly I thought, how I got Table 2 from Table 1: "In order to test the efficacy of the Mayfield model, I took a sample from the population shown in Table 1. In the simulation, I pretended that I 'discovered' the 40 nests on different days by removing the X's from Table 1 prior to arbitrarily chosen days of discovery (Table 2)." I just do not see how this could be made clearer for the reviewer, short of my holding his hand as we typed the X's etc. into his spreadsheet.

What does the reviewer mean, "If the deletion was done in some repeatable manner and could be done many, many times, it would be very useful to see the overall performance of the estimators called into question"? I don't have a clue what he can be getting at, but it does not represent evidence that I am wrong about the Mayfield method.

Fifth (c), the reviewer writes, "When the data for Table 2 are examined, it is noteworthy that all nests are still observed for at least some days. This again raises the point that it would be useful to better understand how data deletion was accomplished when moving from Table 1 to Table 2." I answered this in 5(b). I don't know why this should be so mysterious to the reviewer, unless, of course, he does not study nests in the field (you know, finding nests and filling out nest record cards). If I am showing the history of a nest between discovery and success or failure, should there not be "at least some days"? I suppose I could have included a nest found on one day and destroyed before the next census, but I did not. It really does not matter (if you understand what is going on).

The reviewer continues, "If one were to randomly choose a day of discovery repeatedly, it would be likely that at least some nests that fail prior to fledging would go undiscovered on at least some simulations." What can this possibly mean? As I understand the real world, a nest can only be discovered once, not repeatedly.

He continues, "This [the quotation in previous paragraph] is the very reason that Mayfield began work on this problem in the first place." Well, actually, Mayfield began work on the problem because he was concerned that when the simple method was used on samples with nests found late in the nesting episode, there were biases that resulted in overestimates of nest success.

On the very first page of his 1961 paper, Mayfield wrote, "To illustrate, suppose we were to find a series of nests on the eve of hatching. In this special group the 'hatching success' would be nearly 100 per cent, and subsequently the 'nest success' and 'success of eggs to fledging' would be almost twice as high as if these nests had been discovered at the start of incubation."

The reviewer continues, "For the purposes of this paper, I think that it is very important to show that the average or expected performance of an estimator is flawed. Simply showing that a single realization from an un-described data-deletion process results in estimates that don't match the underlying estimate for the whole population is not compelling to me." I can understand why the reviewer may not find my arguments compelling. He has no idea of what is going on with regard to estimating nest success from field data.

The reviewer lectures me, "Realize that when the data from Table 2 are analyzed with a time-varying survival model in Program MARK that the resulting estimates simply mimic exactly what is provided for daily binomial trials in Table 2. That is, daily survival rates for the 2 days are estimated as: [he gives a big table of statistics here]." The reviewer interprets his table, "these estimates can be multiplied together to estimate the probability of surviving all 12 days (0.2160611; 95% CI=0.113 to 0.375). It is true that these estimates don't match the daily estimates obtained from Table 1 [He's right about that!]. But it is also true that these are based on a sample of the data provided in Table 1. [So what? The numbers are wrong!]"

Day	DSR (reviewer)	DSR (my Table 1)
1	0.9375	0.9750
2	0.8000	0.9231
3	0.9333	0.9722
4	0.9286	0.9714
5	1.0000	1.0000
6	0.9333	0.9706
7	0.8125	0.9091
8	0.9286	0.9667
9	0.7619	0.8276
10	0.8261	0.8333
11	0.8000	0.8000
12	0.9375	0.9375
Nest success	0.2161	0.3750

Wow! What can I say? The true data for the population under consideration, in which we observed all nests from day of first-egg-laying through survival or failure, are found in my Table 1. Thus, the data in the right hand column are the true daily survival rates. The reviewer's DSR (center column) are determined with "a time-varying survival model in Program MARK." This is what I mean by ornithologists not having a "numbers sense." The reviewer runs my data through a black box and concludes that these are the best estimates (with standard errors yet) for daily survival rates of this population. (No explanation of how he got them except by running some unstated numbers through a black box.) There is not even a glimmer of doubt in the reviewer's mind that this exercise constitutes a test of the "time-varying survival model in Program MARK," which has failed miserably. For him, computer programs tell the unchallengeable truth. Give me a break!

The reviewer notes, "And, it is noteworthy that the 95% confidence interval just includes the true value of 0.375. So, again, I would be very interested in knowing what happens if you repeatedly sample from the data in Table 1 (along with a description of how you sample from Table 1) in terms of overall performance."

OK. Let's do another simulation. (I do not need to do thousands to find out that my criticisms of the Mayfield method are correct. All I do is open my spreadsheet and change the X's around.) All this requires, really, is a numbers sense. Let's say that the data in my Table 2 were collected by Tom. Let's say that instead of Tom, we had Dick census the population (let's call it Table 3, at the end of my letter). This is called a "thought experiment" by Einstein and others. One must imagine in one's brain what is going on in the field.

In this example, all I did was change the date of discovery of the nests in this sample. The days of success or failure are the same as in Table 1 and Table 2. This results in fewer observation days between day of discovery and day of success or failure (256) than in Table 2 of my paper. Therefore, the daily mortality rate (DMR) = 25/256 = 0.098, DSR = 1 − DMR = 0.902, nest success = 0.902^{12} = 0.291, which is different from 0.1997 from the sample of this population in Table 2. Nevertheless, the actual nest success of this population is still 0.375 (= 25/40). I do not need a thousand repetitions to tell me what is going on.

TABLE 3. Simulated spreadsheet of a sample of 40 nests, each of which was monitored from day of first egg laying to day of failure (F) or success (S)[a] for days 0-12/

Nest	Day of nest episode (x^2)													Days[b]
	0	1	2	3	4	5	6	7	8	9	10	11	12	
1								X	X	X	X	X	S	5
2	X	X	X	X	X	X	X	X	X	F				9
3							X	X	X	X	X	X	S	6
4		X	F											1
5	X	X	X	X	X	X	X	X	X	F				9
6							X	X	X	X	X	F		5
7	X	X	X	X	X	X	X	X	X	X	X	X	S	12
8	X	X	X	X	X	X	X	F						7
9				X	X	X	X	F						4
10					X	X	X	X	X	X	X	F		7
11						X	X	X	X	X	X	X	S	7
12										X	X	F		2
13					X	X	X	X	X	X	X	F		7
14			X	X	X	X	X	X	X	X	X	X	S	10
15	X	X	X	X	X	X	X	F						7
16	X	X	X	X	X	X	X	X	X	F				9
17	X	F												1
18				X	X	X	X	X	X	X	X	X	S	9
19								X	X	X	F			3
20	X	X	F											2
21				X	X	X	X	X	X	X	X	X	S	9
22							X	X	X	X	X	X	S	6
23	X	X	X	X	X	X	X	X	X	X	F			10
24					X	X	X	X	X	X	X	X	S	8
25										X	F			1
26						X	X	X	X	X	X	X	S	7
27					X	X	X	X	X	X	X	X	F	8
28	X	X	X	X	X	X	X	X	X	X	X	X	S	12
29	X	X	X	X	X	X	X	X	X	X	F			10

	0	1	2	3	4	5	6	7	8	9	10	11	12	
30	X	X	X	X	F									4
31		X	F											1
32	X	X	X	X	X	X	F							6
33									X	X	X	X	S	4
34				X	X	X	X	X	F					5
35					X	X	X	X	X	X	X	X	S	8
36							X	X	X	X	X	X	S	6
37				X	X	X	X	X	X	F				6
38	X	X	X	X	X	X	X	X	X	F				9
39		X	X	F										2
40	X	X	X	X	X	X	X	X	X	X	X	X	S	12
N_x^e	15	17	15	19	23	25	28	27	27	24	20	16	15	
Total														256

[a] The nest episodes have been truncated to 13 days [twelve 24-hour intervals] in order to save space.

[b] Day 0 is the day on which the first egg was laid in the nest. X indicates the presence of eggs or chicks in the nest. Young fledge on day 12 (the 13th day of a nest episode). **S** indicates successful departure of at least one chick. **F** indicates nest failure (no chicks known to have fledged).

[c] Exposure days for each nest.

[d] Survival is the probability of surviving from day 0 to day x and is given by $\prod_0^x d_x$ $(d_x = N_{x+1}/N_x)$.

[e] Total numbers range from 15 nests at day 0 to 15 at day 12.

So, now you have three samplings from the population shown in Table 1 of my paper.

Number of exposure days	Clutch success
199	0.1997
256	0.2914
370	0.4319

Look at that! The greater the number of exposure days, the greater the estimate of clutch success, just as stated in my paper and just as explained in Point 3, item 3B above. This stuff is extraordinarily elementary. What can I do to explain the rather simple mathematical issues involved? How many different ways do I have to explain this without getting, "Bert Murray is wrong; the correct answers can be found in Shaffer's (or somebody's) black box."

Summary of my comments to this reviewer:

What on earth can I say? I guess I have written enough for you to get the idea that I am exasperated because your reviewers do not know the nature of the problem— calculating nest success—and evidently cannot do the rather simple mathematics (all of which I learned by the time I was a Freshman in high school). It seems a shame that ornithologists are going to continue miscalculating clutch success or, as they seem to prefer, daily survival rates. Someday, all of this work is going to have to be redone because, sooner or later, there will show up among the ranks of ornithologists some thinkers who not only remember their high school math but are respected.

Cheers,

Bert

Selected Bibliography

Jehl, J. R., Jr., and B. G. Murray, Jr. 1986. The evolution of normal and reverse sexual size dimorphism in shorebirds and other birds. Current Ornithology, 3: 1-86.

Jehl, J. R., Jr., and B. G. Murray, Jr. 1989. Evolution of sexual size dimorphism: a reply to Mueller. Auk 106(1): 155-157.

Murray, B. G., Jr. 1979. *Population Dynamics: Alternative Model.* Academic Press, New York.

Murray, B. G., Jr. 1982. On the meaning of density dependence. Oecologia, 53: 370-373.

Murray, B. G., Jr. 1984. A demographic theory on the evolution of mating systems as exemplified by birds. Pp. 71-140 in Evolutionary Biology, vol. 18 (M. K. Hecht, B. Wallace, and G. T. Prance, eds.). Plenum, New York.

Murray, B. G., Jr. 1985a. Evolution of clutch size in tropical species of birds. Pp. 505-519 in *Neotropical Ornithology* (P. A. Buckley, M. S. Foster, E. S. Morton, R. S. Ridgely, and F. G. Buckley, eds.). Ornithological Monographs, 36. Lawrence, Kansas.

Murray, B. G., Jr. 1985b. The influence of demography on the evolution of monogamy. Pp. 100-107 in *Avian Monogamy* (P. A. Gowaty and D. Mock, eds). Ornithological Monographs, 37. Lawrence, Kansas.

Murray, B. G., Jr. 1985c. Population growth rate as a measure of individual fitness. Oikos, 44(3): 509-511.

Murray, B. G., Jr. 1987. The calculation of r of populations with heterogeneous life histories. Oikos 50(2): 262-266.

Murray, B. G., Jr. 1988. On measuring individual fitness: a reply to Nur. Oikos 51(2): 249-250.

Murray, B. G., Jr. 1990. Population dynamics, genetic change, and the measurement of fitness. Oikos 59: 189-199.

Murray, B. G., Jr. 1991a. Sir Isaac Newton and the evolution of clutch size: a defense of the hypothetico-deductive method in ecology and evolutionary biology. Pp. 143-180 in *Beyond Belief: Randomness, Prediction, and Explanation in Science* (J. Casti and A. Karlqvist, eds.), CRC Press, Boca Raton, FL.

Murray, B. G., Jr. 1991b. Measuring annual reproductive success, with comments on the evolution of reproductive behavior. Auk, 108(4): 942-952.

Murray, B. G., Jr. 1991c. Comments on the use of Lotka's discrete equations. Oikos, 62(1): 118-122.

Murray, B. G., Jr. 1992a. The evolutionary significance of lifetime reproductive success. Auk, 109(1): 167-172.

Murray, B. G., Jr. 1992b. The evolution of clutch size: a response to Wootton, Young, and Winkler. Evolution 46(5): 1581-1584.

Murray, B. G., Jr. 1994a. Effect of selection for successful reproduction on hatching synchrony and asynchrony. Auk 111(4): 806-813.

Murray, B. G., Jr. 1994b. On density dependence. Oikos 69(3): 520-523.

Murray, B. G., Jr. 1995a. On the evolution of mating systems: a comment on Arnold and Duvall. Am. Nat. 146(3): 475-478.

Murray, B. G., Jr. 1995b. A method for projecting genotypic change in populations with complex genetic and demographic structure. Oikos 73(3): 415-418.

Murray, B. G., Jr. 1997a. On calculating birth and death rates. Oikos, 78(2): 384-387.

Murray, B. G., Jr. 1997b. Population dynamics of evolutionary change: demographic parameters as indicators of fitness. Theor. Popul. Biol. 51(3): 180-184.

Murray, B. G., Jr. 1999a. Predicting the occurrence of synchronous and asynchronous hatching in birds. Pp. 624-637 in Proc. 22 Int. Ornithol. Congr. (N. J. Adams and R. H. Slotow, eds.). Durban, South Africa.

Murray, B. G., Jr. 1999b. Is theoretical ecology a science?: a reply to Turchin (1999). Oikos 87(3): 594-600.

Murray, B. G., Jr. 1999c. Can the population regulation controversy be buried and forgotten? Oikos 84(1): 148-152.

Murray, B. G., Jr. 1999d. Is theoretical ecology a science?: a reply to Turchin (1999). Oikos 87(3): 594-600.

Murray, B. G., Jr. 2000a. Universal laws and predictive theory in ecology and evolution. Oikos 89(2): 403-408.

Murray, B. G., Jr. 2000b. Measuring annual reproductive success in birds. Condor, 102: 470-473.

Murray, B. G., Jr. 2000c. Dynamics of an age-structured population drawn from a random numbers table. Austral Ecology, 25(4): 297-304.

Murray, B. G., Jr. 2000d. Density dependence: reply to Tyre and Tenhumberg. Austral Ecology 25(4): 308-310.

Murray, B. G., Jr. 2001. Are ecological and evolutionary theories scientific? Biol. Rev. 76(2): 255-289.

Murray, B. G., Jr. 2003. A new equation relating population size and demographic parameters: some ecological implications. Ann. Zool. Fennici 40: 465-472.

Murray, B. G. Jr. 2005. Erratum. Ann. Zool. Fennici 42:

Murray, B. G., Jr. 2006. A new equation for calculating reproductive success of clutches as a function of the day on which incubation starts: some implications. Auk 123(3):708-721.

Murray, B. G., Jr., and V. Nolan, Jr. 2007. A more informative method for analyzing reproductive success. Journal of Field Ornithology 78(4): 401-406

Murray, B. G., Jr., and L. Gårding. 1984. On the meaning of parameter x of Lotka's discrete equations. Oikos, 42(3): 323-326.Murray, B. G., Jr., and V. Nolan Jr. 1989. The evolution of clutch size. I. An equation for predicting clutch size. Evolution 43(8): 1699-1705.

Murray, B. G., Jr., J. W. Fitzpatrick, and G. E. Woolfenden. 1989. The evolution of clutch size. II. A test of the Murray-Nolan equation. Evolution 43(8): 1707-1711.

Chapter 7

Karl Popper, Laws, and Science

[On 12 February 2008 (and earlier) the home page of the Web Site for the journal *Oikos* stated that among the journal's most accessed papers by Synergy users in 2005 were two Opinion pieces, "The anarchist's guide to ecological theory, Or, we don't need no stinkin' laws," by R. B. O'Hara, and "Ecological laws: what would they be and why would it matter?," by Marc Lange. On 7 September 2005, I submitted a critical commentary on these two papers. Inasmuch as this commentary was rejected (on 28 October 2005), readers of these papers were deprived of having an opportunity of reading a dissenting view. The manuscript follows as it was submitted. Following is my commentary on the reviews that provide the reasons for the editor's rejection of my paper. My further comments on what I have written in both the paper and commentary on reviews are foot-noted.]

Karl Popper, Laws, and Science

Popper (1989: 19) advises us never to take "seriously problems about words and their meanings." Nevertheless, philosophers (e.g., Lange 2005) and biostatisticians (O'Hara 2005) continue to debate the meaning of "law" and its usefulness in biology. These debates distract us from what Popper (1989: 19) suggests is important, "What must be taken seriously are questions of fact, and assertions about facts: theories and hypotheses; the problems they solve; and the problems they raise." What I think Popper meant is that regardless of whether we call Newton's statement about inertia a law, principle, hypothesis, premise, assumption, conjecture, or guess, Newton's theory remains unchanged.

Thus, scientists should have been investigating (as they were) the usefulness of Newton's statements in predicting and explaining the observable physical world, rather than discussing the meaning of "law." They asked, Do Newton's statements make testable predictions? Are these predictions consistent with the empirical facts?

Nevertheless, we humans define terms to facilitate communication. Both Lange (2005) and O'Hara (2005) are concerned with what a law of nature is. I do not argue with Lange, O'Hara, and others regarding their definitions of law. A problem arises, however, when we fail to distinguish among definitions. The danger lies in the participants in the discussion "talking past each other" (O'Hara 2005), as I think we are. Inasmuch as their definitions are not the same as my definition, communication between us and, more importantly, among our readers is certainly impaired.

My definition of law is, *A statement that is presumed to be universally true and from which predictions (statements that can be shown to be true or false) may be deduced.* Popper did not explicitly define "law," but he did write (Popper 1935; English translation, 1968: 59, his italics), "To give a *causal explanation* of an event means to deduce a statement which describes it, using as premises of the deduction one or more *universal laws*, together with certain singular statements, the *initial conditions*." Clearly, Popper meant that his universal laws were intended to be premises of a deductive argument. Furthermore, according to Copi (1953), "Given certain premisses, any conclusion which can logically be inferred from them is regarded as being explained by them. And given a fact to be explained, we say that we have found an explanation for it when we have found a set of premisses from which it can be logically inferred." Thus, my statement (Murray 1979, 1985, 1991, 1999a, 2000, 2001), *Selection favors those females that lay as few eggs or bear as few young as are consistent with replacement because they have the highest probability of surviving to breed again, their young have the highest probability of surviving to breed, or both,* is a law by this definition, even if not a law by other definitions. It is a law because I assumed it to be universally true and I deduced, among other things, a prediction that was unknown to other biologists—that clutch size in birds should be correlated with the length of the breeding season rather than with latitude. A survey shows that tropical species of birds that have large clutch sizes in fact have short breeding seasons (Murray, unpubl.).

Neither O'Hara (2005) nor Lange (2005) refers to Popper or his deductive philosophy of science. Nor does either refer to my candidates for laws in ecology and evolution (Murray 1979, 1985, 1991, 1999a, 2000, 2001). According to Popper, the issue for scientists should be whether biological laws predict any statements that can be shown to be true or false, not whether they qualify as a "law" by some person's definition. If my statement, given above, cannot be falsified by deducing predictions that are contrary to fact, then we cannot *know* that it is untrue.

O'Hara (2005) distinguishes between "correlative" laws, which describe observed regularities, such as Rapaport's law ("diversity increases towards the equator") and "causal" laws, such as Newton's Laws of Motion, which are "intended as descriptions of the way the universe works." The first is a generalization derived from "a large number of observations." The second, according to O'Hara, "has to be true because of the way the world is." O'Hara, however, does not explain how we are to *know* how the world really is. Lange (2005) also distinguishes between two kinds of law— laws of nature, which are statements that are not only universally "true" but "necessary in a scientifically significant sense," such as "all bodies travel no faster than the speed of light," and "accidental laws," which just happen to be universally true, such as "all solid gold cubes are smaller than a cubic mile." Lange, however, found that "no sharp distinction can be drawn between laws and accidents." How can we tell them apart?

Lange (2005: 397) suggests, "g is a law exactly when g belongs to a stable set," which follows from the argument, "Take a logically closed set of truths that is neither the empty set nor the set of all truths. Call such a set stable exactly when every member g of the set would still have been true had p been the case, for every counterfactual supposition p that is logically consistent with every member of the set." Lange, however, does not tell us how we biologists are to recognize and to *agree* on a logically closed and stable set of truths. We ecologists seem unable to agree on whether there are *any* statements at all that we may regard as universally "true" (Lawton 1999, Murray 1999b, 2000, Turchin 2001, Berryman 2003, Colyvan and Ginzburg 2003, O'Hara 2005).

I too distinguish between two kinds of generalization in science. These differ in their origin, universality, and consequences (Murray 1991, 2001, 2004). First, we have statements that are derived as generalizations from a large number of observations (these are the "regularities" of O'Hara 2005). For example, we study species A and discover that individuals of high latitude populations are larger in size than those of closely related low latitude populations. We study species B, C, and D and find the same relationship. Eventually, a biologist notices the regularity and describes what we now call Bergmann's Rule. The reason for calling it a "rule" rather than a "law" is its lack of universality (Rensch 1971). We know of species in which body size in higher latitude populations are smaller than in related low latitude populations. Bergmann's Rule, however, is a perfectly respectable generalization. It describes a large body of examples. It is not a law because it is not universally true, and it does not explain why the individuals of high latitude populations are larger than those of low latitude populations.

This kind of generalization characterizes biological thought. Rensch (1971) identified 100 such correlative statements and considered them rules because most of them were found to have exceptions. The kind of reasoning described here is known among philosophers as "inductive" (Audi 1995, Honderich 1995). Inductive generalizations, then, originate from direct observation and, more often than not, are not universally true. One of the consequences is that they are untestable, even if universally true (Popper 1968). Imagine a world in which *all* 5,000,000 swans are, in fact, white. How could we know that all swans in the world are white? We look at 100 swans, and they are white. We look at 1,000 swans, and they are white. We look at 100,000 swans, and they are white. Are we justified in saying that all 5,000,000 swans in the world are white? No. This is the problem that Hume recognized two centuries ago (Popper 1968, Magee 1973, 1997). Even though all swans are white, we have no way of *knowing* that. We have no justification for believing that swan 1,000,001 will be white even if the first 1,000,000 swans were white. This is the epistemological problem that Hume raised.

Second, there are universal laws, as defined above. Although long practiced by Isaac Newton and other physicists, what has become known as the deductive-nomological (D-N) method was first explicitly described by Popper in 1935 (first published in English in 1959; see quote from 1968 edition in third paragraph above). Popper suggested the D-N method as an alternative to Hume's induction. Explicitly, a D-N theory comprises (a) a set of *universal laws*, (b) a set of *initial conditions*, and (c) the *predictions* deduced from (a) and (b) (Hempel and Oppenheim 1948, Hempel 1965). For example, the laws are conjectures *presumed* to be universally true (e.g., Newton's Laws); the initial conditions are empirical statements thought to be true (e.g., the radius and period of the moon's orbit); and the predictions are statements to be tested (e.g., that a body near the earth's surface falls 16.1 feet in the first second of free-fall, discounting the effects of the atmosphere).

Deductive-nomological theories are deterministic. Given the laws and the initial conditions, certain consequences should be observed; indeed, *must* be observed if the theory is correct. It is the determinism of D-N theories that makes them rigorously testable, and falsifiable. A prediction of Newton's theory is that a body near the earth's surface will fall 16.1 feet, not 160 feet *or* some other distance.[12] Provided the experiment has been carried out properly and the initial conditions met, if the body falls a distance significantly different from 16.1 feet, the theory has been falsified. Although D-N theories are deterministic, the universe they describe is not (Popper 1982). The predictions of a theory are dependent on the initial conditions, which are chosen by humans and which are virtually infinite in number. The questions are, given the proposed laws, which are bold guesses or conjectures about how the universe works (Feynman 1965, Popper 1979) and which cannot be directly verified (Einstein and Infeld 1938), What initial conditions (of the infinite number that exist) are necessary to predict what we can observe (these are the proximate causes)? What are the verifiable consequences of the laws under specified initial conditions?

Inductive rules are based on large numbers of observations. They usually have exceptions, and, therefore, are not universally true. Bergmann's Rule is not "*All* higher latitude populations are characterized by having a larger body size than their related lower latitude populations." It is, "Higher latitude populations *tend* to have a larger body size than related lower latitude populations." Thus, it cannot be refuted, in the sense that the discovery of a species with smaller body size at higher latitudes does not result in its

---

[12] I should have added the words, "in the first second of free-fall."

refutation. The laws of D-N arguments are creations of the human mind (bold conjectures about processes rather than correlated relationships) and are necessarily presumed to be universally true. Newton never observed a body moving in a straight line at constant speed. Nor has anyone else. Scientists accept Newton's laws because they work—that is, their predictions have been repeatedly verified by observation and experiment. Universal laws are tested by predicting their consequences under various combinations of initial conditions. Newton's Laws and "initial condition set A" predict that an apple will fall 16.1 feet in the first second of free-fall near the earth's surface. Newton's Laws and "initial condition set B" predict the orbit of a comet. Newton's Laws and "initial condition set C" predict the existence and position of an otherwise unknown planet. And so on. As long as the predictions are verified by observations, we can say that the laws have not yet been falsified, but we can never conclude that the laws are true. We accept them because they work, not because they are true. Newton's Laws were initially falsified in 1859 when they could not account for the precession of Mercury's orbit. Although Newton's theory has been falsified as an explanation of the physical world, Newton's Laws have not been discarded because they are useful in solving many physical problems.

Whether ecology and evolution have "laws of nature" or not is a matter of definition and not worth arguing (Popper 1989). Whether aspects of the ecology and evolution of living things can be deduced from universally applicable statements, whatever they are called, is another issue. Without belaboring it too much further, I think my colleagues who are interested in this issue about laws in ecology, should attempt to falsify my guess about the evolution of clutch size, given above. As far as I know, it is universally true. Even though we cannot ever *know* it to be universally true (Feynman 1965, Popper 1979), we all can know whether or not it has been falsified. So far, there has not even been an attempt at falsification. Thus, there is no *logical* reason for anyone to believe that it is untrue.

Facts

O'Hara (2005:391) states, without any reference to the physical literature, "Colyvan and Ginzburg [2003] cite the failure of the law of conservation of kinetic energy [because it was said to have exceptions]. But, this has been supplanted by the law of conservation of energy (which in turn has been supplanted by Einstein's theories of relativity!)." Nothing in this statement is true. First, the law of conservation of kinetic energy applies only for perfectly elastic collisions, whereas the law of conservation of energy applies to all (elastic and inelastic) collisions.[13] Thus, the latter did not replace the former. Second, Einstein's theories of relativity have not replaced the physicists' "law of conservation of energy." Einstein's special theory of relativity, however, with its famous equation, $E = mc^2$, showed that mass was another form of energy and resulted in the merging of the then separately known "law of conservation of energy" and "law of conservation of mass" into the "law of conservation of mass-energy" (Rothman 1995).

Although non-physicists (e.g., Colyvan and Ginzburg 2003, O'Hara 2005) tend to believe that physical laws of

---

[13] On further searching, I have been unable to find a reference to the "law of conservation of kinetic energy" in any textbook or popular book on physics. I can find a reference to it only on the Internet. A physicist, whom I asked about the law of conservation of kinetic energy, replied that not only had he never heard of it but it made no sense. When I wrote this sentence, I had accepted what was on the Internet as bearing some relationship to the real world of physicists.

nature have "exceptions," Feynman (1995: 69, his italics) wrote, "There is a fact, or if you wish, a *law*, governing all natural phenomena that we know to date. There is no known exception to this law—it is exact so far as we know. The law is called the *conservation of energy*. It states that there is a certain quantity, which we call energy, that does not change in the manifold changes which nature undergoes."

O'Hara (2005: 390) wrote, again without any reference to the physical literature, "it certainly appears that laws have become old-fashioned—at least for physicists. And if physicists feel that they do not need laws, why should anyone else?" There are many books that discuss the laws of physics. I shall mention one, *The Character of Physical Law* (Feynman 1965). Thus, O'Hara's (2005) subtitle, "we [ecologists] don't need no stinkin' laws," seems to be a conclusion based on a profound misunderstanding of the history and practice of physics.

The need for laws in science was probably expressed best by Einstein (quoted in Barnett 1950), "The grand aim of all science is to cover the greatest number of empirical facts by logical deduction from the smallest number of hypotheses or axioms," and, more specifically, by Popper (1979: 191), ". . . it is the aim of science to find *satisfactory explanations*, of whatever strikes us as being in need of explanation. By an *explanation* (or a causal explanation) is meant a set of statements by which one describes the state of affairs to be explained (the *explicandum*) while the others, the explanatory statements, form the 'explanation' in the narrower sense of the word (the *explicans* of the *explicandum*)." We need laws because laws explain our world by allowing us to predict not only what we already know empirically but that which we do not know (see Copi, quoted above.)

Finally, Dyson (1988: 45-46) described the difference between physicists and biologists, "Unifiers are people whose driving passion is to find general principles which will explain everything. They are happy if they can leave the universe looking a little simpler than they found it. Diversifiers are people whose passion is to explore details. … They are happy if they leave the universe a little more complicated than they found it. … Biology is the natural domain of diversifiers as physics is the domain of unifiers." Does the diversifier philosophy of biologists derive from the nature of biology or from a self-imposed restriction on thought?

Acknowledgement I thank J. R. Jehl, Jr., for his comments on this paper.

References

Audi, R. (ed.). 1995. The Cambridge Dictionary of Philosophy. Cambridge University Press.

Barnett, L. 1950. The meaning of Einstein's new theory. Life 28: 22.

Berryman, A. A. 2003. On principles, laws and theory in population ecology. Oikos 103: 695-701.

Colyvan, M. and Ginzburg, L. R. 2003. Laws of nature and laws of ecology. Oikos 101: 649-653.

Copi, I. M. 1953. Introduction to Logic. Macmillan.

Dyson, F. J. 1988. Infinite in All Directions. Harper & Row.

Einstein, A. and Infeld, L. 1938. The Evolution of Physics: From Early Concepts to Relativity and Quanta. Simon and Schuster.

Feynman, R. P. 1965. The Character of Physical Law. MIT Press.

Feynman, R. P. 1995. Six Easy Pieces: Essentials of Physics Explained by Its Most Brilliant Teacher. Addison-Wesley.

Hempel, C. G. 1965. Aspects of Scientific Explanation and other Essays in the Philosophy of Science. The Free Press.

Hempel, C. G. and Oppenheim, P. 1948. Studies in the logic of explanation. Philosophy of Science 15: 135-175.

Honderich, T. (ed.). 1995. The Oxford Companion to Philosophy. Oxford University Press.

Lange, M. 2005. Ecological laws: what would they be and why would they matter? Oikos 110: 394-403.

Lawton, J. H. 1999. Are there general laws in ecology? Oikos 84: 177-192.

Magee, B. 1973. Popper. - Fontana Press.

Magee, B. 1997. Confessions of a Philosopher. Random House.

Murray, B. G., Jr. 1979. Population Dynamics: Alternative Models. Academic Press.

Murray, B. G., Jr. 1985. Evolution of clutch size in tropical species of birds. Pages 505-519 in Neotropical Ornithology (P. A. Buckley, M. S. Foster, E. S.

Morton, R. S. Ridgely, and F. G. Buckley, eds.). American Ornithologists' Union.

Murray, B. G., Jr. 1991. Sir Isaac Newton and the evolution of clutch size: a defense of the hypothetico-deductive method in ecology and evolutionary biology. Pages 143-180 in Beyond Belief: Randomness, Prediction, and Explanation in Science (J. Casti and A. Karlqvist, eds.). CRC Press.

Murray, B. G., Jr. 1999a. Predicting the occurrence of synchronous and asynchronous hatching in birds. Pages 624-637 in Proceedings of the 22nd International Ornithological Congress (N. J. Adams and R. H. Slotow, eds.), BirdLife South Africa. (Published on a CD.)

Murray, B. G., Jr. 1999b. Is theoretical ecology a science? A reply to Turchin (1999). Oikos 87: 594-600.

Murray, B. G., Jr. 2000. Universal laws and predictive theory in ecology and evolution. Oikos 89: 403-408.

Murray, B. G., Jr. 2001. Are ecological and evolutionary theories scientific? Biological Reviews 76: 255-289.

Murray, B. G., Jr. 2004. Laws, hypotheses, guesses. American Biology Teacher 60: 598-599.

O'Hara, R. B. 2005. The anarchist's guide to ecological theory. Or, we don't need no stinkin' laws. Oikos 110: 390-393.

Popper, K. R. 1935. Logik der Forschung. (Not seen by me.)

Popper, K. R. 1968. The Logic of Scientific Discovery. Harper & Row.

Popper, K. R. 1979. Objective Knowledge: An Evolutionary Approach (revised ed.). Oxford University Press.

Popper, K. R. 1982. The Open Universe: An Argument for Indeterminism. Routledge.

Popper, K. R. 1989. Conjectures and Refutations: the Growth of Scientific Knowledge. Routledge.

Rensch, B. 1971. Biophilosophy. Columbia University Press.

Rothman, T. 1995. Instant Physics. Fawcett Columbine.

Turchin, P. 2001. Does population ecology have general laws? Oikos 94: 17-26.

Commentary (*Oikos*, 12 January 2006)

[Readers may find most of this commentary inordinately boring. The reviewers have made so many errors over and over again that my responses become repetitive. I have written this for the reviewers' edification and for the historical record, giving to future historians of science an explicit example of the reasons for the lack of intellectual progress in ecological and evolutionary theory during the last quarter (perhaps half) of the 20th century and continuing into the 21st century. I am afraid that, after 40 years of frustration with reviewers, the reviews of this paper caused me to become more pointed in my comments regarding the reviewers' intellectual capacities.]

Dear …

With each submission of a manuscript nowadays, the quality of the reviews sinks lower than I can even imagine. The two that I received from *Oikos* on my ms, "Karl Popper, laws, and science" (MS O14736), have got to be approaching the bottom of the barrel of incompetence. On the basis of our prior history, I know that there is little point in my appealing to you for a change in your decision to reject my paper. I submitted the paper because I thought it was a corrective to the misstatements and illogic of Lange and O'Hara and that, therefore, it was a contribution to the discussion on the existence and usefulness of laws in ecology, which has largely taken place in the pages of *Oikos*. I had hoped that you would ignore our personal history and seriously consider it for publication in the interests of our science.

I write the following commentary as a responsible scientist, scholar, and educator with the hope that you will send it on to the reviewers. If they pay attention to what I have written, I think there is the possibility they will learn something.

Furthermore, I take the time to write these long replies because I think that posterity should have a historical record of the development of our science. Theoretical ecology is not developing because of the poor scholarship exhibited by ecologists[14] in what they write anonymously (and even publish, e.g., Lange and O'Hara). I think this reply to the reviews should make interesting reading regarding the state of ecological theory at the beginning of the 21st Century.

My reply is long because the reviewers made an enormous number of errors, most of which had nothing to do with my criticism of the papers by Lange and O'Hara, but affected their understanding of my paper.

First, a general point: I submitted my ms. as an Opinion piece that was commenting specifically on two papers appearing in *Oikos* (110(2), 2005), one by Marc Lange, the other by R. B. O'Hara, both of the papers themselves being Opinion pieces. According to the blurb in *Oikos*,

---

[14] Philosophers and statisticians as well.

"**Opinion** is intended to facilitate communication between reader and author and reader and reader. Comments, viewpoints or suggestions arising from published papers are welcome. Discussion and debate about important issues in ecology, e.g., theory or terminology, may also be included. Contributions should be as precise as possible and references should be kept to a minimum."

Both reviewers criticized me for not having written a magnum opus. From what they have written, it is clear to me that they were unaware that I was writing an Opinion piece and that they had not read the Opinion papers I was commenting on (Lange's and O'Hara's), about which they say nothing, or what I have written before, including my long philosophical paper (*Biological Reviews*).

Reviewer 9409:

The first reviewer is certainly a philosopher. Today's philosophers of science are intent on minimizing science as a way of knowing about the world we live in. In their writings (or at least those that I have been exposed to) they misrepresent what scientists do and even misrepresent each other. I have criticized them severely in my *Biological Reviews* paper. So far, no one has even argued that anything that I have written is factually untrue.

My commentary on this review:

(1) This reviewer wrote (his emphasis) that I should have written "a clear, coherent, comprehensive, critical, but non-partisan and accessible REVIEW for ecologists of how various scholars and schools have treated the nature of laws, and what they think count as specific instances of laws in ecology."

He continued his criticism: According to him, my paper was "1) a very modest account of Popper's view, 2) an inadequate review of selected positions on the nature of laws, and 3) an opportunity for us to hear what Murray thinks (personally, rather than analytically) about various positions on ecological laws. Ultimately, the paper offers little that is new, and does an inadequate job of reviewing what has already been developed."

Continuing (emphasis his), "The paper could be enormously improved by focusing on one of three approaches: 1) review the nature of laws in ecology, 2) review Popper's insights [sic] (and limitations) as they relate to ecological laws, OR 3) rigorously advance (or critique) the development of some particular notion of what an ecological law is."

Continuing, "The topic of 'laws' is sufficiently complex to warrant focusing an entire paper on just one of these."

"Finally, the title of the paper suggests much more than is delivered. I expected a more-or-less thorough, scholarly overview of Popper's views on laws and science and their implications for ecology. Indeed, the paper certainly makes substantial reference to Popper. However the paper is not about Popper, as the title suggests. Instead the paper seems to be to be [sic] much less."

The reviewer does not point to any error of fact or logic in my paper. His complaint is that I did not write the paper that he wanted to read. He seems not to have understood that my paper is a commentary on the Opinion pieces of Lange and O'Hara, neither of which presents a "full review" of the issues this reviewer expected from mine, and both present *their idiosyncratic* views on the nature of scientific laws, completely out of context of what has been written before. The reviewer seems unaware that I have written several papers on the subject, some of which were cited in the references, and most of which were published in *Oikos*. The reviewer is unaware that I have not only written but published a lengthy review of the kind he describes (*Biological Reviews*, 2001). The text was 20,000 words, and

I cited 300 references. I criticized philosophers pretty severely in that paper, and I have received no complaints that I have misrepresented their views. They really should not misrepresent mine. No one has yet pointed out an error of fact or logic in that paper. In any case, I am sure that my *Biological Reviews* paper would not have qualified as an Opinion piece, which was intended to point out the errors of fact and logic perpetrated on the pages of *Oikos* by Lange and O'Hara. The reviewer does not point out that I had written anything wrong in my description of Lange's and O'Hara's contributions.

Although I stated that Popper suggested that we should not argue the definitions of words and that I agreed with him, the reviewer wanted me to provide (I repeat, reviewer's emphasis) "a clear, coherent, comprehensive, critical, but non-partisan and accessible REVIEW for ecologists of how various scholars and schools have treated the nature of laws, and what they think count as specific instances of laws in ecology."

The reviewer suggests that my big review be "non-partisan." I wonder if he has ever read a publication, especially one labeled "Opinion" that was non-partisan. If so, I should like to have a reference to it.

(2) The reviewer not only failed to understand the point of my Opinion, he failed to understand what I wrote. He states, "the paper seems importantly to be a platform for:

1) Murray to assert what he thinks a law ought to be."

"Ought!? OUGHT!? The *first* sentence of my paper is, "Popper (1989: 19) advises us never to take 'seriously problems about words and their meanings.'" The *first* sentence of the second paragraph states, "Nevertheless, we humans define terms to facilitate communication." How does the reviewer get the idea that I am proposing what *ought* to be the definition of "law"? According to my dictionary, "ought" is "used to express duty or moral obligation," "used to express justice, moral rightness, or the like," "used to express propriety, appropriateness, etc.," "used to express probability or natural consequence." Nowhere have I ever suggested what the definition of law *ought* to be. As a proposer of laws in ecology, what I did was to define "law" so that others might know what I meant by the term when I used it (there seem to be many other definitions of "law" in the literature)

I have elsewhere pointed out that Humpty-Dumpty stated rather clearly, "When *I* use a word, it means just what I choose it to mean—neither more nor less," to which Alice replied (her emphasis), "The question is whether you *can* make words mean so many things." "The question is," said Humpty-Dumpty, "which is to be master—that's all." Now, if I claim that I can define a word in any way I wish, then I must concede that anyone else can define the same word any way he wishes. My definition is neither right nor wrong because it disagrees with others. Other definitions are neither right nor wrong because they disagree with mine. I am not the final arbiter on the meaning of words, and, as I have already said, nowhere have I even hinted at the possibility that my definition of the word "law" *ought* to be used by anyone else, much less by *everyone* else as a moral obligation. As I stated in my ms., rather explicitly I thought, I defined the word for the purpose of facilitating communication between me and my readers. My use of "law" is not inconsistent with my definition. No one can criticize my thoughts about clutch size on the basis of someone else's definition of law. One may point out that my proposed laws in ecology and evolution are not consistent with other definitions, such as Lange's, and that's OK with me—I will not argue the point. We have to decide instead whether we want to argue, as scientists, about the definition of law or about the predictions that follow from my proposed universal statement (whatever it is called).

(3) The reviewer states (my italics), "The assertion [he is referring to my definition] is not accompanied by an analysis supporting his assertion, a clear connection to what others (who have expressed them deeply) think a law *ought* to be, or an adequate analysis of the implications for Murray's sense of a law."

This reviewer must be a philosopher (who else talks about the "analysis" of the meanings of words?). As such, he is incapable of understanding what I am writing about because I am a scientist interested in thinking of theories that explain the world that I am living in. In the first place he suggests that I discuss what others (allegedly deep thinkers on the subject) think that a law *ought* to be. As I pointed out in the first sentence of my paper (it seems necessary for me to remind my critics of what I have written), "Popper (1989: 19) advises us never to take 'seriously problems about words and their meanings.'" Why should I even be interested in writing about what others think about the meanings of words? I can understand the intense dislike that philosophers who are devoted to the analysis of the meanings of words have toward Popper. Popper clearly implies that what they think is worth doing is just not worth doing. A moment's reflection indicates that Popper is right.

The reviewer states that I have not provided an "analysis" (whatever that may mean to a philosopher) supporting my so-called assertion. Let's get the words straight: what the reviewer calls my assertion is actually my definition (there is a difference in meaning between "assertion" and "definition"), as I stated in the text. Furthermore, I pointed out what one "deep thinker," Karl Popper, wrote about the relationship between universal laws and their logically deduced predictions. Contemporary philosophers seem to have nothing but contempt for Popper, and, as far as I can determine, these philosophers have not a clue about what Popper in fact thinks. My judgment on this is based on what philosophers have in fact written. What scholars must do, if they are interested in these issues, is compare what Popper has written with what has been written about him. I criticized philosophers pretty severely in my *Biological Reviews* paper. So far, no one has pointed out that I am wrong.

I wrote to Marc Lange, indicating that I was puzzled about why he had not cited Popper. He replied that Popper's view, regarding the rejection of induction and support for deduction, "as on so much else, is widely repudiated—by philosophers, explicitly, and by scientists, implicitly." Lange wrote further (I have indicated Lange's direct quotation of Popper with italics), "[Popper] says that a statement is a law *'if and only if it is deducible from a statement function which is satisfied in all worlds that differ from our world, if at all, only with respect to initial conditions.'*" Lange continued, "The problem should be obvious: His definition of 'law' presupposes the concept of an 'initial condition,' i.e., a contingent truth that is not a law, and so is circular. He might as well just have said 'A law is a truth that is not dependent on any truth that isn't a law.'"

My problem with Lange's observation on Popper is that I could not find the sentence quoted by Lange. What I found instead was (Popper's italics), "a statement may be said to be logically necessary if and only if it is deducible . . . from a *'universally valid'* statement function; that is to say, from a statement function which is *satisfied by every model.* (This means, true in all possible worlds.)" Lange disrespects Popper so much that he could not even quote him correctly. Note that the words "initial conditions," which appear in Lange's quotation of Popper, do not appear in Popper's sentence at all. Note that Popper does not define "law" but a "logically necessary statement," which is a *prediction* of a deductive-nomological theory—a prediction that *follows* from the laws. To tell the truth, I thought that Lange was going to reply by pointing out that I had found the wrong quotation, but, alas, he did not.

Note also that Lange's misquotation of Popper indicates that a law is *"deducible from a statement function,"* but, from everything I have read, laws are a priori statements from which other statements (i.e., the predictions of the theory) are deduced. (I point out further that initial conditions are not contingent truths. They are, more often than not, statements of fact, such as the period and radius of the moon's orbit.) Lange's misquotation of Popper results from his not understanding what a law is in a deductive argument. Furthermore, a law does not *depend* on anything—it is an a priori (i.e., axiomatic) statement.

This reminds me of a statement by philosopher of biology David Hull, "philosophers are prone to scandalous carelessness in transcribing quotations and to inaccurate descriptions, not to mention some highly questionable interpretations."

Why are philosophers so incompetent? I do not know. If any were to reply, I am sure that they would undertake an analysis of the meaning of "incompetent." In any case, Lange did not write back to me why he had misquoted Popper.

Who can have respect for such sloppy scholarship? Can scholars (including me) be mistaken in interpreting someone's writing? Of course, but blatant misquotations?

[To be fair, what I believe happened was that Lange copied the quote from someone else rather than from Popper directly. Lange did not deliberately misquote Popper. Thus, Lange's views on Popper are based on misinformation obtained second hand from other philosophers, who, in order to discredit Popper and his philosophy, invented the misquotation. That, however, does not say much for Lange's scholarship.]

(4)   The reviewer wrote, "Murray to express disappointment that neither O'Hara nor Lange refer [sic] to Popper or Murray. I'm not sure that they ought to have."

The first rather convoluted group of words (not a sentence) requires me to guess what the reviewer means. I think he meant that I was disappointed that neither O'Hara nor Lange referred to either Popper or me. He is right. He offers his opinion that neither O'Hara nor Lange should have mentioned our publications. That just simply reveals the reviewer's bias. It happens that Popper and I (and others, such as Einstein and Feynman) represent an alternative to O'Hara's and Lange's rather idiosyncratic views about science. There happens to be many sides to the issue, and Popper's, Einstein's, Feynman's, and my side are under-represented in recent years. By leaving the view of Einstein, Feynman, Popper, and me out of their discussions, did either Lange or O'Hara provide their readers with "a clear, coherent, comprehensive, critical, but non-partisan and accessible REVIEW for ecologists of how various scholars and schools have treated the nature of laws, and what they think count as specific instances of laws in ecology."?

What the reviewer has not done is show that anything that I have written is incorrect, either factually or logically. It is hard for me to understand why the editors of a scholarly journal should want to suppress one side of the issue—a side that has not been shown to be false in any way. To be sure, there are many who disagree with my views, but I am puzzled by why they do not want to address in public the issues that I raise. The reviewers may think that their arguments are solid, inasmuch as they are protected by anonymity and cannot be held accountable for their positions.

(Minor point. This reviewer makes the same grammatical mistake that the second reviewer made regarding the verb form that goes with "neither O'Hara nor Lange ...." It should be "refers," not "refer.")

(5) The reviewer wrote, "Ultimately, the paper offers little that is new, and does an inadequate job of reviewing what has already been developed."

Well, what does it mean that I had presented nothing new? What I did was seriously question the idiosyncratic views of O'Hara and Lange, which were published in *Oikos*. I also posed a simple question. If my statement, which I happen to call a law (i.e., *Selection favors those females that lay as few eggs or bear as few young as are consistent with replacement because they have the highest probability of surviving to breed again, their young have the highest probability of surviving to breed, or both*) leads to predictions of biological facts hitherto unknown to biologists, should it not be of interest to biologists (and even philosophers), regardless of whether we call it a law or a guess? Would it not serve as a counterexample to the widespread *belief* among ecologists and philosophers of science that biology, or even science, can have no universal statements that serve as laws in a deductive-nomological theory?

I wonder why alleged scholars want to debate the meaning of the word "law" rather than discuss the deeper issues raised by the existence of universal (unverifiable but falsifiable) statements that lead to verifiable singular statements.

Reviewer 12523

Editors' note: Murray's frustrated response to this reviewer spanned 53 pages. He knew it would fall on deaf editor ears but, as a teacher, hoped it might be sent on, so that the reviewer might "wake up and smell the roses." In it he re-argued his case and commented on the educational system, how ecologist should go about doing science, the philosophy of science, the peer review system, the editors judgment in selecting the reviewer, and the near-impossibility of publishing dissenting views. As for the review itself, nearly 40% of the reviewer's 51 numbered comments (on a 12 page manuscript, mind you) dealt with grammar and syntax.

When it came to substance and ideas, the reviewer could offer neither refutations nor notations of error. The best he could do was repeat "I'm not sure..." Would you be outraged at such treatment? A selection of Murray's comments follows.

This reviewer is the most intellectually incompetent person that has come my way in the past 50+ years. He seems to know nothing about the history and philosophy of science and seems also to be uninterested in biology as a science. He is a product of the slipshod educational system of the last quarter of the 20th Century, in which the goal of education has become making students "feel good about themselves" rather than to educate them I do not subscribe to this educational philosophy. In his review, he repeatedly admits he is "not sure of this or that, e.g., whether Popper or Einstein, or whoever, really meant what I was attributing to them (usually with quotations!) Because he is "not sure" of anything, he is convinced that I am wrong. He criticizes me for thinking like Newton, Einstein, Feynman, and Popper. He also questions the meaning of virtually every word and sentence that I wrote, including my spelling, grammar, and syntax (even though English is probably not his first language).

My advice to him (actually, to all reviewers):

(1) Never criticize someone unless you are on solid ground: Your criticisms, which begin, "I'm not sure about this or that," just do not suffice. When you criticize someone, provide supporting facts, reasons, arguments, and references. Substantiate your position as you would in a paper meant for publication.

(2) Criticize only "errors" of fact or logic. That is where the reviewer's expertise is supposed to be.

(3) Never criticize an author's use of language, unless you are recommending the paper for publication and are trying to be helpful. There is no point whatsoever for you to criticize the language in a paper you are recommending for rejection (other than to show off—and you can show off only when your criticisms are correct).

(4) I recommend that you never criticize an author's use of language unless it is your first language.

(5) Always sign your name to a negative review. Be accountable for what you think. Be professional.

Because the reviewer does not understand anything about my paper, let me give another quotation from Einstein, which may help him to comprehend the context of what I have written:

"The truly great advances in our understanding of nature originated in a way almost diametrically opposed to induction. The intuitive grasp of the essentials of a large complex of facts leads the scientist to the postulation of a hypothetical basic law, or several such laws. From these laws, he derives his conclusions. . . which can then be compared to experience. Basic laws (axioms) and conclusions together form what is called a 'theory.' Every expert knows that the greatest advances in natural science... originated in this manner, and that their basis has this hypothetical character."

Why is it that what "every expert knows," according to Einstein, is not known by ecologists? I just assume that they do not know any better. But why are they and the philosophers so resistant to what I have to say, especially since what physicists have to say about their approach to understanding nature is so explicit and clearly contradicts what biologists and philosophers believe about them.. I can only conclude that ecologists just do not know how physicists think about their problems.

Now we may proceed with commentary on the review.

This review is probably the worst review that any of my papers over the past 45 years has received. It serves as an example of everything wrong with "peer review."

The last thing I expected when I submitted my ms. was to receive an English lesson from a reviewer, much less from a reviewer for whom English is a second language. The reviewer has little to say about the content of my paper, so he has padded his review with commentary on my spelling, grammar, and syntax. About 20 of his 51 comments refer to English usage. He was right on one or two of them.

A second problem with the review is that the reviewer was "not sure" of this and "not sure" of that. The reviewer is honest enough to say that he is "not sure" or "not positive" about many things, but if he is not sure, why is he commenting on what I have written (almost all with references to my sources)? Why did he not read my references? In any case, does the fact that the reviewer is unsure about what I have written make what I have written wrong? The reviewer's admitted lack of knowledge on the subject should not be accepted as evidence that I am *necessarily* wrong. I have highlighted, in bold, the reviewer's uncertainty (in over 12 of his 51 comments).

Reviewers today seem not to understand that their responsibility is to be helpful to both the editor and the author. A reviewer should find and *document* errors of fact or logic committed by an author, not offer unsubstantiated opinion. This reviewer, like the last, did not identify any errors of fact or logic in my paper, much less document them.

Editors' note: Below are selected excerpts.

I suggested that physicists, instead of debating Newton's description of his assumptions as "laws," asked, "Do Newton's statements make testable predictions? Are these predictions consistent with the empirical facts?"

The reviewer asks, "Why aren't these questions about usefulness? The author asserts that these are questions, instead, about laws. What would questions about usefulness look like then if not like the questions they asked?"

The questions that I ask here are clearly about the usefulness of Newton's laws. Why should the reviewer think otherwise? After all, didn't I write (emphasis added), "Thus, scientists should have been investigating (as they were) the *usefulness* of Newton's statements in predicting and explaining the observable physical world, rather than in discussing the meaning of 'law'?" The usefulness of a theory in science is in how well it predicts what we can observe and, therefore, explains the empirical facts. Anyone concerned with the usefulness of a theory must then ask the very questions that I suggested they ask. Scientists did not concern themselves with the meaning of Newton's choice of the word "law" to describe the assumptions of his theory. This, however, is the concern of O'Hara and others, who would like to deny that biology or even science can have statements called laws. A statement can be accepted as a law only if it has predictable consequences that can be compared with empirical facts (that's Einstein's, Feynman's, Popper's, and my view, anyway). Nevertheless, Lange, O'Hara, and others do not ask whether my laws in fact are useful in predicting testable consequences. If my laws work, then their philosophies of science turn out to be so much bunk. Thus, I think my critics, if they are scientists, should be asking questions about the usefulness of my laws for predicting biological facts. If my critics are philosophers (in the old sense of the term—i.e., persons interested in understanding the world they live in rather than in the meaning of words), then they too should answer the questions that I pose here. If my laws do not work, then my theory is so much bunk. Why do we not have an open discussion.

***

I wrote, "Clearly, Popper meant that his universal laws were intended to be premises of a deductive argument."

The reviewer wrote, "**It's not clear to me** that this is what Popper meant – why does the author think it is clear?"

So what did Popper write? We are *not* discussing my paraphrase or interpretation of what I think Popper meant. I *quoted* him. I *quote* again, "To give a *causal explanation* of an event means to deduce a statement which describes it, using as premises of the deduction one or more *universal laws*, together with certain singular statements, the *initial conditions*."

I think that Popper clearly meant that a causal explanation comprises a set of premises that include (i) one or more universal statements, called laws, and (ii) initial conditions. From these one deduces a statement that describes an observation—such as, an apple should fall 16.1 feet in the first second of free-fall near the earth's surface.

I will give the reviewer another quotation on this point from Popper (his italics),

"… it is the aim of science to find *satisfactory explanations*, of whatever strikes us as being in need of explanation. By an *explanation* (or a causal explanation) is meant a set of statements by which one describes the state of affairs to be explained (the *explicandum*) while the others, the explanatory statements, form the 'explanation' in the narrower sense of the word (the *explicans* of the *explicandum*)."

What Popper describes here is the deductive-nomological model of Hempel and Oppenheim (1948), which is sometimes called the Hempel-Popper model. This is all treated in some detail in my previous publications, which I recommend that the reviewer read. I would also recommend that he read some of Popper and the Hempel/Oppenheim paper. Because the reviewer is unfamiliar with this literature and the ideas, he is certain to have problems understanding what is going on because it is not easy. As a scientist, when writing a scientific paper, I am writing for my so-called peers. If the reviewer's interpretation of Popper is different from mine, he should not write, "It's not clear to me that this is what Popper meant." Instead, he should clearly explain what he thought Popper meant, so that we may judge whether he understood Popper. The evidence that he does understand at all what Popper wrote is lacking. He also does not seem to understand me. Therefore, I am justified in thinking that he does not understand the subject very well.

***

I wrote (original italics), "If my statement, given above [that is, *Selection favors those females that lay as few eggs or bear as few young as are consistent with replacement because they have the highest probability of surviving to breed again, their young have the highest probability of surviving to breed, or both*], cannot be falsified by deducing predictions that are contrary to fact, then we cannot *know* that it is untrue."

The reviewer wrote, "**I'm not sure** that this is saying anything. So what if we cannot know that it is untrue. A lot of things are like that – infinite number of things perhaps, but what does that do for us? Perhaps nothing?"

This comment makes me wonder whether this reviewer is a scientist at all! What are we scientists to be doing if not searching for the truth (emphasis on searching)

266

as expressed in unfalsified universal statements? That is what Einstein, Feynman, and Popper tell us that scientists should be doing. Does the reviewer believe otherwise? Are scientists supposed to be searching for untruths? Anyone can write an untrue statement. If the reviewer is correct, then perhaps we ecologists should state clearly that scientists value untruth over truth. Maybe we should encourage the publication of untrue statements. Untrue statements or statements without supporting evidence are of no interest to me.

It seems rather obvious to me that if no one can demonstrate that my statement is untrue (by deducing a prediction that is contrary to fact), then no one can *know* that my statement is untrue...

The real problem with this reviewer's comment is that he is willing to believe statements that may be untrue. He says he does not care whether a statement is true or not. Unfortunately, people's actions are based on what they believe to be true, not on what happens to be true. This is one reason for caring whether a statement is true or not. That is why we have to sort them out, and that is what science is about. As Feynman stated, "That principle, the separation of the true from the false by experiment or experience, that principle and the resultant body of knowledge which is consistent with that principle, that is science." (I am sure this reviewer will be trying to sort out Feynman's mangled syntax rather than his meaning. That's too bad. What Feynman is trying to say is more important than the manner in which he has phrased it.) If separating what is true from what is false is not science, then what is science?

***

I wrote, "We ecologists seem unable to agree on whether there are *any* statements at all that we may regard as universally 'true' (Lawton 1999, Murray 1999b, 2000,

Turchin 2001, Berryman 2003, Colyvan and Ginzburg 2003, O'Hara 2005)."

The reviewer wrote, "Is this a problem? Why or why not? Or, does this create an opportunity for a new and unique thought about the nature of ecology and ecological explanations? This seems ripe for development into something interesting."

Yes, it is a problem. If the concept of laws in scientific arguments is rejected out of hand for philosophical reasons, then we cannot even discuss a proposed law (such as mine) on scientific grounds. Ecologists and philosophers do not test my proposed law because they believe, a priori, that biology can have no laws. That is a philosophical position, and that is a problem of the first magnitude for ecology (and evolution).

*****

I wrote, "It [Bergmann's Rule] is not a law because it is not universally true, and it does not explain why the individuals of high latitude populations are larger than those of low latitude populations."

"The reviewer wrote, "**I'm not sure** why a law has to both be universally true <u>and</u> explain why something is universally true – doesn't this raise the bar a bit too high or expect laws to do something that they really aren't setting out to do? Even the author's own definition or notion of law above doesn't ask for this – where did it come from?"

The reviewer is not sure again. First of all, a law does not "explain why something is universally true"—a law explains *why* a particular fact or event (i.e., the prediction) occurs. This is the meaning of the quotation from Copi (1953), quoted above, but which the reviewer could not understand the point of my quoting…

It also comes from the preceding Popper quotation in the ms., which I repeat (Popper's italics), "To give a *causal explanation* of an event means to deduce a statement which describes it, using as premises of the deduction one or more *universal laws*, together with certain singular statements, the *initial conditions*."

The reviewer seems to think that I am suggesting that "laws … do something that they really aren't setting out to do." Well, of course, laws are not really setting out to do anything. Scientists, however, who develop deductive-nomological theories are certainly intending to *explain* empirical facts with their laws.

The reviewer should learn that one reads words, sentences, and paragraphs in the context in which they are written. One sentence or paragraph sets the stage for later sentences and paragraphs.

***

I wrote, "The kind of reasoning described here is known among philosophers as 'inductive' (Audi 1995, Honderich 1995). Inductive generalizations, then, originate from direct observation and, more often than not, are not universally true. One of the consequences is that they are untestable, even if universally true (Popper 1968)."

The reviewer wrote, "**I'm not sure** about this – **I'm not sure** that this is how philosophers talk about induction. It seems somewhat backward. Moreover [sic], **I'm not sure** that it is true that inductions are 'untestable' as the author claims in the next sentence. I can make an inductive claim that all golf balls have dimples based on 1 million observations, <u>and</u> at the same time continue to test that claim. Also, the point of inductions is that I cannot know if they are universally true, hence they are inductions.

If the reviewer really wants to know how at least some philosophers talk about induction, he should read a few philosophers on the subject, perhaps beginning with those I cited. I recommend especially that he read **induction** in *The*

Cambridge Dictionary of Philosophy (cited in my paper) and Popper (1968). I am distressed that I must instruct one of my alleged peers on philosophical problems that have been known for 200 years.

The reviewer, again, states that he is not sure about something. In that case, perhaps he should not comment. Or, perhaps, he should not comment until he has read at least my references to philosophers who have written about induction. For example, the reviewer wrote, "Though this may (or may not) be Popper's claim, it's far from clear that it is an accurate depiction of induction." Why does he not *know* whether it's Popper's claim or not? Is he too lazy to look it up? Why does his ignorance of what Popper and other philosophers have written about induction mean that what I say about what Popper said about induction is wrong?

***

In the context of this whole discussion by the reviewer, I invite him to read the following passages (from Bryan Magee, *Confessions of a Philosopher* [his italics]):

"It was in relation to the philosophy of science that Popper worked out his most fundamental ideas: that we are never able to establish for certain the truth of any unrestrictedly general statement about the world, and therefore of any scientific law or any scientific theory (it is important to be clear that he is talking not about singular statements [e.g., an apple will fall 16.1 feet in the first second of free-fall near the earth's surface] but about unrestrictedly general ones [e.g., Newton's First Law of Motion]: it is possible sometimes to be sure of a direct observation, but not of the explanatory framework that explains it: direct observations and singular statements are always susceptible of more than one interpretation); that because it is logically impossible ever to establish the truth of a theory, any attempt to do so is *an attempt to do the logically impossible*, so not only must logical positivism be abandoned because of its verificationism but also all philosophy and all science involving the pursuit of certainty must be abandoned, a pursuit which had dominated Western thinking from Descartes to Russell; that because we do not, and never can in the traditional sense of the word "know," know the truth of any of our science, all our scientific knowledge is, and will always remain, fallible and corrigible; that it does not grow, as for hundreds of years people believed that it did, by the perpetual addition of new certainties to the body of existing ones, but by the repeated overthrow of existing theories by better theories, which is to say chiefly theories that explain more or yield more accurate predictions; that we must expect these better theories in their turn to be replaced one day by better theories still; and that the process will have no end; so what we call our knowledge can only ever be our theories; that our theories are the products of our minds; that we are free to invent any theories whatsoever, but before any such theory can be accepted as knowledge it has to be shown to be preferable to whatever theory or theories it would replace if we accepted it; that such preference can be established only by stringent testing; that although tests cannot establish the truth of a theory they can establish its falsity—or show up flaws in it—and therefore, although we can never have grounds for believing in the truth of a theory, we can have decisive grounds for preferring one theory to another; that therefore the rational way to behave is to base our choices and decisions on "the best of our knowledge" while at the same time seeking its replacement by something better; so if we want to make progress we should not fight to the death for existing theories but welcome criticism of them and let our theories die in our stead."

And:

"The address [of Popper at Oxford] was called 'Back to the Pre-Socratics' and appears under that title in Popper's subsequent book *Conjectures and Refutations*, published in 1963. Its main contention is that the only practicable way of expanding human knowledge is by an unending feedback process of criticism. Put like that it might seem self-evident, but the real clout of the thesis lies in what it denies. It denies that we get far if we attempt to base the extension of our knowledge on observation and experiment. Observations and experiments, it contends, play the same role as critical arguments; that is to say, they may be used to test theories, challenge theories, even refute theories, but are only ever relevant in so far as they constitute potential criticisms of theories. The way we add to our knowledge is by thinking up plausible explanations of hitherto unexplained phenomena, or possible solutions to problems, and then testing these to see if they fit or work. We subject them to critical examination, try them out on other people to see if anyone can point out flaws in them, devise observations or experiments that will expose any errors they may contain. The logic of the situation is this: we start with a problem—it can be practical, but need not, it can be purely theoretical, something we wish to understand or explain; then we use our understanding of the problem plus our powers of insight and imagination to come up with a possible solution; at this stage our possible solution is a theory which might be true and might be false, but has hitherto not been tested; so we then submit this conjecture to tests, both the tests of critical discussion and the tests of observation and experiment—all of which, if they are to be tests at all, must constitute potential refutations of the theory. Hence the title *Conjectures and Refutations*, which encapsulates a whole epistemology."

This is about as clear an exposition of Popper's philosophy of science ever written. (It may be worth knowing that Magee had discussed philosophy with Popper almost weekly for almost fifty years.)

I have read a lot over the past many years. Popper and Magee are not idiosyncratic. For example, I recently reread K. C. Cole (*The Universe and the Teacup: The*

Mathematics of Truth and Beauty), a science writer (neither a philosopher nor a scientist), who wrote,

"It was Galileo who abandoned the more cerebral, theory-based truths and decided that the ultimate arbiter should be experiment. Experiments test whether or not theories are true. Every time the experiment turns out the way theory says it should, the theory gains a degree of validity; it gets more and more true. But it cannot ever become completely "true" unless the scientist performs an infinite number of experiments. That holds the door open for revisions based on new revelations. Laws of nature can be proven untrue, but rarely true. Scientific truths are always provisional."

The *only* places in which I read about a contrary epistemology of science are in the ecological and the philosophical literature. I just do not understand why ecologists have not yet bought into this philosophy of science (I understand why philosophers don't). Perhaps someone will one day take the time to explain it to me.

***

I wrote, "Are we justified in saying that all 5,000,000 swans in the world are white? No"

The reviewer wrote, "This has not been established. We may well be justified in making such a claim or in making inductive leaps (how else would we be able to live in the world?). The real question is at what point are we justified in making this claim, or are there no points of justifications [sic], just greater or lesser certainty?"

If the reviewer had read *anything* about inductive logic, he would know that it has been established for at least 200 years that we humans have no logical way in which to know that all swans in the world are white, even if they were. He may *believe* that they are all white, but he cannot *know* that they are all white. This is simple logic. And, I repeat, it has been *established* for at least 200 years. I am not making this stuff up.

***

The reviewer does not have to believe that Popper's and Copi's views regarding the relationship between prediction and explanation are correct. That is certainly OK. It's a philosophical issue. Unfortunately, the reviewer has not read any philosophy, so he cannot follow the argument. He has no basis for thinking otherwise. He is not intellectually equipped to evaluate the argument one way or another, much less comment on my thinking. There is an alternative philosophical view now (the anti-Popperians) that contends that you can accept "explanations" even if you are unable to predict anything with your explanation. But this is not how science is supposed to work. Explanations without empirical evidence in the form of predictions are based on faith. The reviewer's lack of understanding on this point does not seem a sufficient basis for rejecting the views of Popper, Copi, and other philosophers and scientists.

The reviewer should be aware that there is a school of philosophy that is bitterly antiscientific. They make no such claim, of course, but they are. They disparage the works of Popper, Hempel, and others. The problem for philosophers is that their status in society has declined with the rise in the status of post-1900 science. Prior to 1900 philosophers were the sages to whom the movers and shakers (e.g., kings and politicians) turned to for advice. That is no longer true (Indeed, at some universities, philosophers are fighting to save their departments). If a government wants to build a bomb, they ask scientists and engineers to build it. They do not even bother to ask philosophers (who are busy debating the meanings of words) whether it is ethical to drop the bomb on civilians. That is too bad for philosophers and philosophy, and, unfortunately, that is too bad for society as a whole. While I believe that philosophers have much to contribute, they have truly marginalized themselves since the mid-20th Century. People who are doers are really not interested in debates about the meanings of words.

***

I wrote, "[Bergmann's Rule] is, 'Higher latitude populations *tend* to have a larger body size than related lower latitude populations.' Thus, it cannot be refuted, in the sense that the discovery of a species with smaller body size at higher latitudes does not result in its refutation."

The reviewer wrote, "No, but that doesn't mean that it cannot be refuted. Evidence to the contrary over time that resulted in an inability to make the generalization would refute it."

Well, in a sense, the reviewer is correct. Over the next few hundred years, biologists collecting data on birds and mammals may well discover that 51% of species with wide latitudinal ranges show that the lower latitude populations have larger body size than the higher latitude species. At that time, Bergmann's Rule would be "refuted." If that is the way that the reviewer intends to establish or test useful "truths" in biology, then we are certainly going to have to be patient. I, for one, do not want to wait that long, especially when we can learn much more about our world and do it faster by thinking of and testing deductive-nomological theories. That is why I have adopted the deductive method in my research. In fact, I have proposed a set of laws that allow us (with appropriate initial conditions) to predict patterns of clutch-size variations, mating relationships, body size dimorphism (in birds and mammals [and perhaps other groups—no tests yet]), that is, if ccologists want to have a theory that predicts "the greatest number of empirical facts by logical deduction from the smallest number of hypotheses or axioms," which Einstein suggests is the "grand aim of all science." (We do not have to wait around endlessly collecting empirical facts, Bacon-like, in the hope of rejecting many of the current explanations of clutch-size variations that exist.) If the reviewer is not in that paradigm of science, then he cannot appreciate my laws. Unfortunately, that is too bad for me and other ecologists and evolutionary biologists who may aspire to the same goal as Einstein and other physicists in having a predictive theory. The reviewer's ignorance is standing in the way.

***

As I have noted before ecologists are very sensitive to being criticized, which amounts to anyone expressing thoughts contrary to their own. So they describe me as "confrontational" or worse and my writing as "distasteful." The reviewer, however, is correct about my being confrontational. If you think of anything new in science, you are necessarily being confrontational when you express it publicly. I am challenging my critics to show that anything that I have written is false. Inasmuch as they refuse to do that, I must count them as non-scientists.

***

I wrote, "Even though we cannot ever know it to be universally true (Feynman 1965, Popper 1979), we all can know whether or not it has been falsified. So far, there has not even been an attempt at falsification. Thus, there is no *logical* reason for anyone to believe that it is untrue."

The reviewer wrote, "**I'm not sure** that it works this way. I can imagine many claims that someone might make that we have not attempted to falsify but that we might have very good and logical reasons to believe to be untrue. No one has ever attempted to demonstrate that my claim that gremlins exist in my computer, but we still have good reasons to believe that such a thing is untrue. The more interesting question is, then, why would we prima facie believe some things to be untrue while we prima facie believe others to not be untrue prior to attempted falsification?"

First, let us note that the reviewer has never published in a scientific journal that there are gremlins in his computer. If he had, the first response would be to ask for his scientific evidence. If he has none, then we would dismiss him as delusional. There is no reason for any of us to spend our time disproving statements that are lacking evidence. Scientists are not devoting their time to disproving religious myths. They are concerned with finding evidence for or against reasoned hypotheses explaining their observations of nature.

The reviewer asks, "why would we prima facie believe some things to be untrue while we prima facie believe others to not be untrue prior to attempted falsification?"

The answer is simple. Scientists should never *accept* a statement as true without evidence, that is, without failed attempts at falsification. We propose certain statements to be "true" only for the purpose of falsifying them by logically deducing predictions from them.

***

I quoted a passage from Einstein regarding the aim of science: "The grand aim of all science is to cover the greatest number of empirical facts by logical deduction from the smallest number of hypotheses or axioms."

The reviewer wrote, "**I'm not sure** that this is an expression of a need for laws so much as it is a summation of the notion, and the role, of parsimony."

The reviewer focuses his attention on "the smallest number of hypotheses or axioms" and concludes that Einstein is making a plea for parsimony. I focus on "cover[ing] the greatest number of empirical facts *by logical deduction from the smallest number of hypotheses or axioms.*" It is clear that Einstein is suggesting that the aim of science is to include a greater range of empirical facts being explained by (i.e., predicted by) a smaller set of hypotheses. So magnetism, electricity, light, etc., once treated as separate subjects, were one by one eventually combined into a single theory of quantum electrodynamics. Ecologists are not even attempting to cover the greatest number of empirical facts by logical deduction from the smallest number of hypotheses or axioms. Ad hoc hypotheses proliferate. That is another reason for thinking that theoretical ecology is not science.

***

I quoted a passage from Popper regarding the aim of science: ". . . it is the aim of science to find *satisfactory explanations*, of whatever strikes us as being in need of explanation. By an *explanation* (or a causal explanation) is meant a set of statements by which one describes the state of affairs to be explained (the *explicandum*) while the others, the explanatory statements, form the 'explanation' in the narrower sense of the word (the *explicans* of the *explicandum*)."

The reviewer commented, "Again, **I'm not sure** that Popper's comment is about the necessity of laws."

So, the reviewer is not sure. If the reviewer doubts that the quotation from Popper is not about the necessity of laws, what does the reviewer think that this passage is about? I will give him a clue. Inasmuch as the reviewer has not read either Popper or me or apparently anyone else regarding the nature of deductive theory, he is unfamiliar with the meanings of the words. Popper says, in this passage, that it is the aim of science to find satisfactory explanations and that a satisfactory, causal explanation requires what he calls the *explicans*, which comprises a set of universal laws and a set of initial conditions...

Am I supposed to give ecologists a minicourse in philosophy of science whenever I write about philosophy in response to other philosophical papers published in biological journals?

278

***

I am truly very disappointed in these reviews. They show that the quality of ecologists is declining rather than improving. These two reviews and the papers by Lange, O'Hara, Colyvan and Ginzburg, and Wilson (among others) indicate that ecologists and philosophers are terribly confused about the nature of physics and Popper's hypothetico-deductive theory. I would not feel so certain about this if these authors had not made so many simple mistakes. We are not discussing interpretation. I am referring to mistakes in fact and logic. The reviewers' views, the subject of this letter, are completely idiosyncratic. What is really too bad is that we dissenting ecologists no longer have a venue for presenting our position.

It was certainly not my intent to submit another paper to *Oikos*. The errors of fact and logic of Lange and O'Hara, however, were so bad that they deserved to be criticized. Perhaps I am the only ecologist capable of doing that. Unless someone can discover and substantiate an error in my writing, I think that it is too bad for theoretical ecology in particular and ecology in general that I cannot publish my corrections to the reviewers' errors.

Cheers,

Bert

Chapter 8

Mass and Weight
Can Ecologists Read English?

[Karl Popper pointed out that the "works which take notice of my ideas usually ascribe views to me which I have never held, or criticized me on the basis of straightforward misunderstandings or misreadings, or with invalid arguments." I can make the same claim for myself. I have no doubt that my critics attribute to me views that I do not hold—indeed, never held. Their criticisms are based on straightforward misunderstandings, misreadings, and invalid arguments. Is this a characteristic of scholarship? If so, is it a characteristic that scholars should be proud of? By my standards, it is not.

Popper was not alone. With a broader brush, philosopher David Hull wrote, "philosophers are prone to scandalous carelessness in transcribing quotations and to inaccurate descriptions, not to mention some highly questionable interpretations." This is also true of biologists (see Murray, *Biological Reviews* 2001). Again, is this the way that scholars should behave? Anyone can write nonsense, but where are the editors whose responsibility (I think) is to prevent the publication of trash.

In January 2008 I published in the *Auk* (125:232-233) a letter to the editor in which I challenged the two-decades-long practice of ornithologists to describe the amount of stuff that birds are made of as "mass," instead of "weight." This practice seems to have originated with Chardine (1986; Auk 103:832), who published an opinion piece in the *Auk*. I say opinion piece because Chardine, a biologist, did not cite anyone with a physical background in support of his opinion. In my letter, I recommended that biologists use the term "weight" instead of "mass," citing several physicists in support of my view.

In correspondence with the editor of the *Auk*, I identified some points for him to consider: (1) I had asked two physicists independently the question, "Imagine that this saucer is a balance. I place my pen on it, and the readout says 10 grams. Have I measured the pen's weight or its mass?" Both physicists, in separate conversations with me, immediately replied, "Weight." (2) One reviewer of my ms. wrote, "When we look at a piece of metal stamped '1 kg', it means that we have here a mass of 1 kg whose weight (on the Earth's surface) is 1 kg weight or simply 1 kg." In other words, "weight" is measured in metric units, which is remarkably similar to what Rogers (1960, *Physics for the Inquiring Mind*) had to say in my cited and quoted reference. (3) This reviewer also wrote, "Dr. Murray's proposal … "Whereas Dr. Chardine conclusion is at variance with accepted international practice." I presumed these were the reasons for the editor's accepting my paper for publication.

Subsequently, however, the editor accepted and published another letter on the subject (Lidicker 2008, Auk 125:7441). I was not even apprised of the existence of this letter until I saw it in print. This letter, too, like that of Chardine, is an opinion piece. Lidicker, a biologist, did not refer to anyone with a physical background in support of his own position on the issue. Worse, he totally mangled my position. Therefore, on 27 August 2008, I submitted a short correction, which the editor rejected in September 2008.

In the following pages I include (A) my manuscript criticizing Lidicker's published paper, (B) my response to the reviews of that ms., and (C) the cover letter in my resubmission of the ms. to the new editor of the *Auk*.]

MANUSCRIPT

Mass or weight: a reply to Lidicker (2008).— Lidicker (2008:744) stated that "the campaign to persuade us to express weight in newtons (Murray 2008) is not going to be successful anytime soon." Nowhere in my note do I even suggest that ornithologists should use the term "newtons" instead of "grams" to describe the weight of birds, much less launch a "campaign" for ornithologists to do so. In fact, I wrote, "I recommend that we use the correct term, 'weight,' instead of 'mass,' even if we continue (as everyone else does) to use the incorrect (i.e., 'bad') units, kilograms (or kilograms-weight), instead of newtons." Clearly, I would like ornithologists to use the word "weight" instead of "mass" to describe the bulk of birds. I would like them to express weight in grams. I would like them to follow the recommendation of the physicists at Great Britain's National Physical Laboratory (i.e., "The most simple method of weighing is to simply place a test piece on a mass balance and take the displayed reading as its weight" [Davidson et al. 2004:4]). I would like ornithologists to do what everyone else is doing, as suggested by a reviewer of my ms., "Dr. Murray's proposal is what is in practice in law, by international agreement, in science and is historically based and is the practice of the market place." The reviewer added, "Whereas Dr. Chardine is technically correct in what he says concerning mass and weight his conclusion is at variance with accepted international practice." If ornithologists continue to refer to a bird's "mass" rather than to its "weight," I should like them to justify doing so by something other than a reference to their own opinion.—BERTRAM G. MURRAY, JR., *Population Dynamics Research, 249 Berger Street, Somerset, New Jersey 08873, USA. E-mail: bmurray@rci.rutgers.edu.*

Literature Cited

Davidson, S., M. Perkin, and M. Buckley. 2004. The Measurement of Mass and Weight. Measurement Good Practice Guide, no. 71. National Physical Laboratory, Teddington, United Kingdom.

Lidicker, W. Z., Jr. 2008. Mass or weight: Response to Murray (2008) and Chardine (2008). Auk, 125:744.

Murray, B. G., Jr. 2008. Mass or weight: What is measured and what should be reported? Auk 125:232-233.

Commentary (*Auk*, September 2008, and subsequent submission letter)

Dear …:

I wonder why it is that editors think that everyone should have a free shot at me. I am a legitimate target, regardless of what I say. Every time I write criticism of someone else, my criticism goes directly to that person. Et,[15] criticism of me I first see in print? Why was I not asked to review the recent letter by Lidicker (Auk 125:744)?

(1) Lidicker writes, "I submit that the campaign to persuade us to express weight in newtons (Murray 2008) is not going to be successful anytime soon." My first question is, Did Lidicker even read my note? Nowhere in my note do I even suggest that ornithologists should use the term "newtons" instead of "mass" to describe the weight of birds, much less launch a "campaign" for ornithologists to do so. In fact, I wrote, "I recommend that we use the correct term, 'weight,' instead of 'mass,' even if we continue (as everyone else does) to use the incorrect (i.e., "bad") units, kilograms (or kilograms-weight), instead of newtons."

How could anyone with the smallest comprehension of the English language claim that I am urging ornithologists

---

[15] Evidently, a typo. What was intended, I do not know.

to express weight in newtons? I want them to express weight in grams! If they continue to refer to a bird's mass rather than to its weight, I want them to justify doing so by something other than a reference to their own opinion.

(2) My position is based on, among other things, the recommendation of the National Physics Laboratory of Great Britain (I quote from my letter), "'The most simple method of weighing is to simply place a test piece on a mass balance and take the displayed reading as its weight' (Davidson et al. 2004:4)."

(3) Note that Lidicker, like Chardine, did not cite any evidence from the physical literature on the differences between "mass" and "weight." His musings on the subject, like Chardine's, are without substance. Thus, we can now say (paraphrasing from my note), "When deciding between 'mass' and 'weight' to describe the size of birds, ornithologists seem to have a choice between the advice of physicists (cited above) or the unsubstantiated opinions of Chardine (1986) and Lidicker (2008)."

(4) With regard to Lidicker's suggestion that we are actually measuring mass when we use an "old-fashioned two-pan balance" to weigh birds. This is a bit tricky because we are in fact comparing the weights of objects on both pans. Because the force of gravity cancels out of the equation, we have what Rogers calls an "indirect measure" of mass. (See comment below.)

(5) What I could not include in my note were several relevant facts:

Fact 1: On 23 May [2007] I spoke with two physicists separately. To each I posed the following question: "Imagine that this saucer is a balance. I place my pen on it, and the readout says 10 grams. Have I measured the pen's weight or its mass?" Both physicists, in separate conversations with me, immediately replied, "Weight."

Fact 2: The first reviewer of my manuscript, who (I believe) must have been a metrologist on the basis of what he wrote, does not explicitly state, "Bert Murray is right in everything he has written," he does state (my italics), *"Dr. Murray's proposal is what is in practice in law, by international agreement, in science and is historically based and is the practice of the market place."* That seems pretty close to, "Bert Murray is right in everything he has written." The first reviewer did not point to any errors in my note or even propose improvements. Ornithologists, however, seem to think they are special and can go against international practice. My guess is that ornithologists have no intellectual basis for doing what they are doing.

Fact 3: The first reviewer of my manuscript wrote, "Whereas Dr. Chardine is technically correct in what he says concerning mass and weight *his conclusion is at variance with accepted international practice"* (my italics). The reviewer, again, certainly seems to agree with me.

Fact 4: The first reviewer of my manuscript pointed out, as Lidicker has, that with a double-pan balance, one can determine the mass of something by placing the unknown on the left-hand pan and adding known masses to the right-hand pan until both sides balance (indicated by a needle on the rocker-arm). Those of us who have taken quantitative analysis in chemistry have done this many times (in class). What is actually happening is that we are comparing *directly* the *weights* of the stuff on the two pans. Because the gravitational force on both pans is the same, we may cancel gravitation out of the respective force equations, and we can say that we are comparing masses. For physicists this is the *indirect* method for calculating the mass of an object (see Rogers).

Ornithologists usually do not carry double-pan balances and a box of weights into the field with them to weigh birds (but David Lack did when I was with him). With regard to the double-pan balance, Klein [1974: 90, my italics] wrote: "For thousands of years the balance, or two-pan scale, was used to compare the *weight* of the object to be bought or sold and the standard *weights* that should just balance it."

Fact 5: A seemingly acceptable measure (at least physicists use the term even if they consider it "bad") of the force of gravity is "kilogram-weight" (or "kilogram-force"), which was used by Rogers, whom I quoted in my ms., and by the first reviewer. I must admit that I have read a dozen web pages in addition to two cited books (Klein and Rogers) plus my college textbook and a more recent college textbook, and found the discussion quite inconsistent, if not confusing. It took me a little time to figure out what was going on. I am glad that the three physicists seem to be in agreement with me regarding what I am measuring with a balance.

C. Resubmission of manuscript to the new editor of the *Auk* upon his election.

2 December 2009

Dear …:

In 1986 Chardine published a note in the Auk arguing that ornithologists should use "mass" instead of "weight" when reporting the size of birds in grams. Chardine did not provide any supporting evidence, such as a reference to the literature, much less to the physical literature. I never did like the argument. (I thought it was quite arbitrary at the time, and when I protested, Alan Brush insisted that I use "mass" instead of weight if I wanted to have my paper published in the Auk.) I did not get around seriously to looking for the evidence until recently. From what I could determine, physicists prefer the use of "weight" to "mass" for seemingly inappropriate situations, such as reporting the size of birds in grams. I found, for example, the following (S. Davidson, M. Perkin, and M. Buckley, "Measurement good practice guide No. 71: The measurement of mass and weight (2004)," published by Great Britain's National Physical Laboratory): "The most simple method of weighing is to simply place a test piece on a mass balance and take the displayed reading as its weight." That and other references convinced me that the correct word for describing the size of a bird in grams is "weight."

I wrote a letter to the editor, which [was] published in the Auk in 2008. One reviewer of that letter wrote, "Dr. Murray's proposal is what is in practice in law, by international agreement, in science and is historically based and is the practice of the market place." The reviewer added, "Whereas Dr. Chardine is technically correct in what he says concerning mass and weight his conclusion is at variance with accepted international practice." I certainly took that as strong support for my argument.

[The editor] subsequently published a "response" by W. Z. Lidicker, Jr. (Auk, 125:744), in which Lidicker completely misrepresented what I wrote. Lidicker added more nonsense on measuring the bulk of a bird. He did so without a single reference to the physical or metrological literature (whereas I had cited four in support of my argument). [The editor] did not give me an opportunity to see Lidicker's ms., which was a direct personal shot at me, before his accepting it for publication (I did not know of it until I saw it printed in the journal).

I wrote a short reply, which [the editor] refused to publish. Inasmuch as Lidicker's note totally misrepresented what I wrote and is quite wrong in its conclusions, I think this was very unprofessional (according to my standards of professionalism). The last word on a subject should not be deliberate obfuscation. The ultimate question is whether there are any references to support anyone's use of "mass" for "weight" for the purpose of measuring the size of birds. Is it possible that ornithologists are misusing a word, which has no support from the remainder of the intellectual community?

I am submitting my letter to you, as an attachment, for your consideration for its publication.

Cheers,

Bert

[I am unable to find any correspondence regarding this letter. This does not mean that there is none.}